Essential Atlas

of Technology

BARRON'S

First edition for the United States, its territories and dependencies, and Canada published 2003 by
Barron's Educational Series, Inc.
© Copyright of English-language translation 2003 by Barron's Educational Series, Inc.

Original title of the book in Spanish: *Atlas básico de Tecnología*
© Copyright 2002 by Parramón Ediciones, S.A.—World Rights
Published by Parromón Ediciones, S.A., Barcelona Spain

Author: Parramón's Editorial Team
Illustrator: Parramón's Editorial Team
Text: Néstor Navarrete

English translation by Eric A. Bye

All inquiries should be addressed to:
Barron's Educational Series, Inc.
250 Wireless Boulevard
Hauppauge, NY 11788
http://www.barronseduc.com

International Standard Book No.: 0-7641-2421-8

Library of Congress Catalog Card No.: *2002107388*

Printed in Spain
9 8 7 6 5 4 3 2 1

PREFACE

The purpose of this *Essential Atlas of Technology* is to present in a simple and instructional form the processes, instruments, and materials that have had the greatest impact on the development of humankind. This is not a history of technology or a catalog of inventions; above all it is a means for the reader to see how certain essential mechanisms function, what principles have inspired the creation of certain essential processes, and how the most up-to-date items are made.

The intention is to present as broad a picture as possible of the different areas into which technological activity has been divided, but we have placed special emphasis on the areas that generally are considered to be the basic pillars of how we do things, including ones of very recent origin that now occupy a significant place in our way of life, such as imaging, electronics, and data processing technologies.

We hope that the reader will see this work as an effective resource and a useful instrument for gaining knowledge of how humans have so far solved the problems involved in survival.

TABLE OF CONTENTS

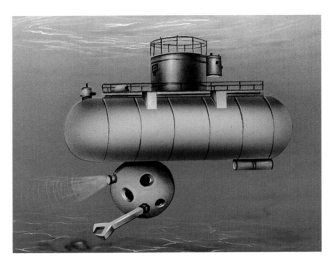

INTRODUCTION

THE TECHNOLOGICAL PRIMATE

Humans are not the only animal capable of modifying natural objects, such as a piece of wood, to convert them into useful tools such as a spear, but our **skill** in doing so surpasses by far that of any other species. In fact, with the passage of time this skill has become the most characteristic behavior of human beings, and even though not all individuals are able to create new ways of producing objects or designing new devices and tools, they are capable of using them after a learning period.

Modern technology is not just a result of the ability to manipulate objects, though; this ability involves another increasingly important factor that is very characteristic of our species: the capacity to **observe phenomena**, sort out their various components, analyze them, formulate hypotheses to explain them, test these hypotheses, and finally, articulate the laws that explain them coherently without contradicting other proven facts, in other words, the ability to **create knowledge**.

The depletion of natural energy reserves makes it imperative to investigate renewable energy sources (solar, wind, and so on). The photo shows wind generators in Denmark.

THE ACCELERATION OF TECHNOLOGY ...

The technological advancement of humanity has not occurred regularly throughout history. For thousands of years our entire technology was restricted to a few stone, wood, and bone implements and some procedures that made daily tasks less arduous for the few nomadic bands that inhabited the earth. **Hundreds of thousands of years** were spent in perfecting stone axes, and about two million years went by between the appearance of the most primitive axes, which were used by an ancestor of *Homo sapiens*, and the first instruments made of metal.

In contrast, barely 60 years passed between the first airplane constructed by the Wright brothers, which was able to fly only a few yards, and the huge turbojets capable of carrying hundreds of passengers across the oceans. This acceleration in technological development is not exclusive to the modern era, but thanks to the application of scientific methods, it has never been as spectacular as at present; according to predictions by experts, it will continue with equal or greater intensity in the coming decades.

Ships, which were invented thousands of years ago, are still very useful in transporting large cargoes.

... AND ITS CONTRADICTIONS

This accelerated development has surely improved the quality of life for many people, but not everyone has benefited equally, and there have also been some undesirable effects. The increasing need to obtain some important raw materials has led to the development of very aggressive extraction techniques that have partly or totally destroyed landscapes and ecosystems, with the resulting loss of valuable plant and animal species.

The railroad has made great contributions to economic development. It requires tremendous infrastructure, such as bridges, tunnels, stations, and so on.

The massive consumption of fossil fuels such as coal and the fuels derived from petroleum is the major cause of **atmospheric pollution**. In addition to endangering the health of many people in some parts of the world and at specific times of the year, it seems to be the main cause of gradual global warming and disastrous climatic changes. The generation of **toxic wastes** that are hard to get rid of and the premature introduction of dangerous processes without adequate safety guarantees are other major problems that the current age of technological acceleration has produced.

Some of these problems are starting to be resolved; others have been clearly diagnosed, and we even know how to deal with them, but putting those means into practice requires political decisions that are sometimes slow in coming. Furthermore, some problems are so new that they are still being studied. Still, in the last 150 years, we have learned that it is not the scientific discoveries or the technological applications that create problems, but rather the **specific economic**, **social**, and **political conditions** under which they are put into practice.

THE INSTRUMENTS

The first steps in what we now recognize as technology occurred with the making of **tools**, and for a long time this was the only technological activity practiced by humans. It was a slow process in which the motivation came from the capacity for observation and the struggle for survival. The most important discovery that stood above all others was the technique for producing fire. In addition to improving food and comfort, mastery of fire made it possible to enhance the production of tools and to discover new **materials** such as terra-cotta to make household equipment and bricks; also, fire ultimately made it possible to **cast metals**, a huge step that led the first settlements that practiced that skill to a previously inconceivable level of development.

There was little more that could be done relying only on intuition, observation, and manual dexterity; however, ancient civilizations had already taken another avenue that would lead them to the first developments that might be considered scientific in nature. The ancient Greeks were the first to apply mathematical knowledge in providing a clear explanation for the function of tools that were already known, such as **levers**, **inclined planes**, and **wheels**. That knowledge made it possible to use calculations to predict how tools and structures would perform before actually making them. It was also shown for the first time that technological advances don't depend exclusively on scientific knowledge, but that economic, social, and political conditions can also be determining factors. The great ancient civilizations had adequate knowledge to develop more efficient techniques and instruments than the ones they had, but the abundance of slaves and the social structure did not make it necessary to do so.

ENERGY

For a long time, humans had only the energy of their **muscles** to use in performing work. The exploitation of **fire** involved the first human victory in the search for outside energy sources to make work easier. The tribes that turned into itinerant shepherds learned to use certain **animals** to transport cargo, and the ones that lived on the banks of rivers and lakes learned to use **wind** to propel rudimentary sailboats; however, most advances occurred when a portion of the population settled down and new needs arose.

Tasks performed by hand included preparing agricultural fields, milling grain to make flour, and transporting water for planted crops, until it was discovered how to use domesticated animals or wind or water power to relieve humans of the hard work involved. Windmills and the paddle wheels moved by flowing water were known to astute mechanics in very ancient times; these devices were created to take advantage of energy sources other than pure muscle power.

Even though there were continued improvements in using these sources of energy, it took several thousand years to discover other sources. It is only 300 years since the first **steam engines** were built, and engines capable of using energy from petroleum and systems for producing electrical energy are advances that sprang from the twentieth century.

THE DESIRE AND THE NEED FOR COMMUNICATION

Travel and communication seem to be two of the primary and strongest impulses of human beings. Technological advances have helped satisfy these desires ever since ancient times. They have provided various advances in **writing** such as terra-cotta tablets, papyrus, and paper, and the broadest assortment of transportation systems such as cars, trains, and airplanes.

But it's only in the last 100 years that technology has been applied so broadly and quickly to transportation and communications and changed the

The quality of digital photographs is approaching that of conventional photography, with the advantage that the images can be manipulated in the home computer.

The CD or compact disc is capable of storing a great quantity of sound and images.

image that our grandparents had of the world. Photography, movies, and television have made us familiar with cities, surroundings, and landscapes located thousands of miles away, and modern airplanes have put them within reach of a few hours travel. Current **communications media** such as television, radio, and the press inform us almost instantly of important news from all parts of the planet, while sophisticated telephone systems make possible almost unlimited communication between individuals.

The creation of the **Internet**, a network that connects millions of computers throughout the entire world, further broadens the possibilities for sharing information immediately and regardless of the distances involved. The aggregate of all these innovations is rapidly changing the way of living and working of millions of people all around the world, but it also gives rise to increasingly important social and economic differences between the rich and poor regions of the planet, and between different social groups within a single country.

THE TECHNOLOGY OF LIFE

The most recent great step in understanding and transforming the conditions in which humans develop has involved **identifying the genes** that define a large number of species, including humans. A species' genome is the sum of biochemical instructions represented by each gene that define the particular traits of that species, from its unique physical appearance to each of the tiny details that govern the function of its organs. The technological possibilities that this discovery presents are enormous, since they run the gamut from **improving**

plant and animal species to make them more productive and resistant to pests and diseases to transforming medical science and making possible such things as preventing and curing hereditary ailments, the creation of **medicines** specifically tailored to each patient, creating **personalized organs** for sick people who need a transplant, and in general, developing a more natural and less aggressive type of medicine.

But in this instance, as well as others already mentioned, the balance between the benefits and the possible drawbacks of applying these complicated and sometimes dangerous techniques will depend on the specific conditions in which they are applied, so the quality of information that we currently have about these subjects becomes very important.

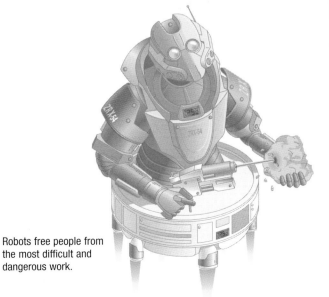

Robots free people from the most difficult and dangerous work.

SIMPLE MACHINES

Some very elementary instruments that are used to **modify forces** are referred to as simple machines. Such simple machines can increase or reduce the force that is applied to them or change the direction or the orientation in which the machines work. Traditionally, the simple machines are considered to be the **lever**, the **pulley** and its derivative the **block and tackle**, and **the windlass**, which work when supported in just one point, plus the **inclined plane**, the **screw**, and the **wedge**, which work through contact on an entire surface. Sometimes the **toothed wheel** and the simple **gear** are added to this group.

LEVERS

Levers serve to increase or decrease the force applied to them. They are divided into three groups according to the location of the fulcrum or support point. If the fulcrum is located between the applied force and the resistance, the lever belongs to the **first type**; when the resistance is located between the applied force and the fulcrum, the lever is of the **second type**; and if the force is applied between the fulcrum and the resistance, the lever is of the **third type**.

DERIVATIVES OF LEVERS

The **scale**, **scissors**, and **pliers** are levers of the **first type**; **wheelbarrows**, **nutcrackers**, and **oars** belong to the **second type**; and **tweezers** are part of the **third type**, which is the only one where the resistance is always less than the applied force.

THE THREE TYPES OF LEVERS

First Type — fulcrum, Roman scale

Second type — wheelbarrow, fulcrum

Third type — tweezers, fulcrum

WINDLASSES AND PULLEYS

Oars are also levers of the second type..

A windlass is a simple machine since it transmits the force directly.

A windlass is nothing more than a cylindrical axle to which a crank or a wheel with a longer radius is added. Windlasses make work easier, because the greater the difference between the radius of the axle and that of the crank, **the less force** is required for such tasks as lifting a weight. The most common image associated with a windlass is the one used with water wells, where a rope from which a bucket is suspended is coiled around an axle when the crank is turned. Pulleys also work with ropes, but if the pulley, which is a wheel with a groove to accommodate the rope, is **immobile**, it does nothing to reduce the effort required; it simply allows for **a better position for pulling on the rope**. In order to multiply the force, the pulley has to be able to **move** along the cord. Usually the two types of pulley are used to make up a **block and tackle**, which makes it possible to achieve both effects.

Three types of pulleys: The fewer movable pulleys there are in the setup, the less force is required to lift a weight.

Fixed Pulley

Movable Pulley

Block and Tackle (fixed pulley plus two movable pulleys)

INCLINED PLANES AND SCREWS

An inclined plane is a ramp that makes it possible to raise a heavy object using less effort than that required in lifting it directly. The **force** that must be applied to a weight to make it go up an inclined plane **decreases** as the **length** of the plane increases.

Although the screws we are accustomed to are the ones that are used to hold two pieces together, these cylinders with a spiral groove that runs their entire length are also machines that make it possible to **change direction of movement**. If a nut is attached to a screw that is turning in place on an axis, the nut moves along the screw and changes from **rotational** to **rectilinear** movement.

As compared to lifting, less effort is required in raising a barrel to a platform when an inclined plane made of planks is used.

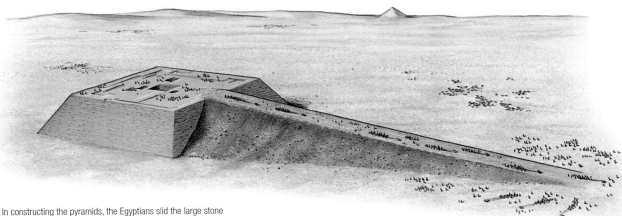
In constructing the pyramids, the Egyptians slid the large stone blocks along an inclined plane.

WEDGES

A wedge is a solid object with an acute angle at one end. When a wedge is applied to an object that it is able to penetrate, the force applied to the wedge along one of the sides of the groove through which it enters is equal to the product of the length of one side of the wedge multiplied by the force it exerts on the object penetrated. That is why the more acute the angle of the wedge, the more the force applied to it is multiplied. Nails are part of the wedge family.

Nails are a type of wedge.

The mathematical explanation of how simple machines work made it possible to use them as a basis for developing complex machines.

SIMPLE TOOLS

The most common tools are often derivations or combinations of simple machines adapted to a specific use by the people who make them. A tool is **any object** that produces a change in some other object on which it is used. Traditional occupations such as agriculture, carpentry, masonry, quarrying, and blacksmithing have developed a multitude of tools, some of which have remained in use for a very long time; others have been modernized. Some tools have come into being more recently as new occupations and materials have been introduced.

FAMILIES OF TOOLS

Some main groups can be distinguished among the thousands of tools in existence; these give rise to hundreds of specialized instruments that work according to the same principles. The tools presented here are some of the most common members of the major families.

The scythe was introduced more than 2,500 years ago and is still in use.

In its forms as a weapon and a tool, the axe dates back to prehistoric times.

STRIKING TOOLS

Instruments that produce their effect by **striking** are among the oldest that have ever been invented. There are two groups that can be identified among them: edged tools and blunt tools. Among the first category the most noteworthy ones include the axe, which has a very long history and currently is used for such tasks as cutting wood, and sickles and scythes, ancient agricultural tools used for harvesting grains and clearing weeds from the fields. Shears, which are used for cutting metals, are a modern offshoot of the axe. There is even more variety in the family of hammers and mallets; they have no sharpened edge, but they exist in a great variety of shapes and sizes, from the tiny hammer used in making jewelry, up to the enormous mallets used in forging iron or in minting coins and medallions.

MOTORIZED AXES AND HAMMERS

Many modern machines are nothing more than enormous axes or hammers operated by powerful motors. That is the case of many punch presses that stamp out materials and are driven by electric or diesel motors, jackhammers that use compressed air to deliver blows, and powerful cutters used on metals, rubber, and plastics.

Jackhammer being used to dig a trench.

CONTACT TOOLS

These tools don't need to deliver blows to accomplish their work, but sometimes it is necessary to strike them with other tools. They belong to the family of **needles**, **punches**, and **knives**. They are used for such varied tasks as **sewing**, **cutting**, **drilling**, **removing material**, **inlaying**, and **polishing**. There are specialized ones for working in metals, wood, cork, leather, and textiles, and as with the percussion tools, the most modern ones are powered by electric or internal combustion motors. The contact tools are generally used for precision work, and as a result, every branch of industry and crafts where they are used has developed a multitude of these tools for specific tasks.

Introduction

Machines and Tools

The Steam Engine

Internal Combustion Engines

Energy

Electricity Production

Electric Motors

Mining

Metallurgy

The Chemical Industry

Construction Materials

Public Works

Transportation Vehicles

Imaging

Electronics

Computer Science

Robotics

Index

SAFETY

You need to be extremely careful in using any kind of tool, especially ones with sharp areas and cutting edges.

Various types of pliers.

Using a chisel to cut a groove.

CUTTING TOOLS

When you think of cutting tools, the first one that comes to mind may be a knife, a razor blade, or the kind of cutter that is used in handwork, but some of the most modern ones also include scissors and pliers. This extended family also includes **chisels** and **gouges**. All of these are long instruments of iron or steel that end in a cutting edge of various widths, and that are used to **remove material**. They almost always have a wooden handle, and they are used in carpentry, quarrying, and metal crafts, even though the tools go by different names in each of these occupations.

SAWS

Wood is made up of long, irregular fibers that are difficult to cut cleanly. With their **toothed cutting edge**, saws are the solution to the problem. The sharp "teeth" of the saw cleanly cut the intricate wood fibers. In addition to **handsaws**, there are circular saws capable of cutting materials as hard as **stone** and **metal**, thanks to the special materials from which they are made.

A backsaw is used to make precise cuts in wood.

A rasp, which is a tool for removing material and polishing, is used to smooth down the saw cut.

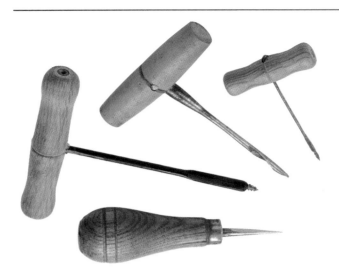

PIERCING TOOLS

In such tasks as sewing, construction, and assembly of metal or wooden parts, one ancillary operation involves the piercing of some material. The first tools used for these purposes were **sharp-pointed**, **needlelike instruments** made from **hard** materials—awls, for example—but with time people discovered tools that required less effort to use. Drills and drill presses are tools that use a twist drill bit to pierce materials by cutting through them and removing the excess material in the form of shavings.

An awl (with the fat handle) can be used to start a hole; gimlets (with transverse handles) have a threaded point and are used to make small holes.

BIT BRACES AND ELECTRIC DRILLS

In order to accomplish their task, drills need to be held in a device that turns them at high speed. A **hand drill** is a tool with a handle and a set of gears that spin the drill at high speed when the operator turns the crank. Electric drills have largely replaced hand drills; they spin faster and provide greater control in drilling. The invention of **mechanical drills**—a combination of electric motor and drill bit—made it possible to drill very hard materials and to make much larger and deeper holes. This in turn required the development of drills made from very hard materials, such as industrial diamonds.

A bit brace is a hand tool that is used to make holes of a specific diameter.

Electric drills make it possible to drill holes quickly, precisely, and easily.

TOOLS FOR STOCK REMOVAL AND POLISHING

In addition to the tools in the chisel family, which were mentioned earlier, there are others specifically designed for removing and polishing material. **Carpenters' planes** create smooth finishes on woods by removing thin shavings in a forward and backward movement, but for a perfectly smooth finish, woods and other materials have to be sanded. For manual polishing, **sandpaper** or **steel wool** is used. Sandpaper is densely coated with very fine **particles of glass** or some other **abrasive material** capable of polishing soft materials. As with previously mentioned cases, most polishing today is done by machines fitted with an electric motor to make the work easier.

A carpenter's plane; the blade can be adjusted to remove more or less material.

Sheets of sandpaper; the finer the grain, the smoother the polish on the treated surface.

UNITS OF MEASURE

For many years the units for the most common measurements such as length, weight, and volume varied from place to place. It was only toward the end of the eighteenth century that a system was devised that would gradually gain universal acceptance: the **decimal metric system**, which is now used in nearly all countries in the world. The tools that are customarily used for measuring are manufactured by using these units as a model and copying them as precisely as possible.

TOOLS FOR MEASURING AND MARKING

The most common tool for measuring length is the tape measure, but there are others such as the Vernier caliper and other types of gauges that serve the same purpose, especially in cases where great precision is required, or when the shape of the object is such that a Vernier caliper can't be used. Many times measurements are taken for the purpose of doing **sketches** or drawings that later are **transferred** to the materials being worked. In such cases tools such as the scale, the set square, and the compass are used.

MEASURING

Measuring involves comparing two quantities of the same type. When the length of something is measured, for example, its length is compared to that of a known **unit** or **model**, such as a **meter** or **yard**.

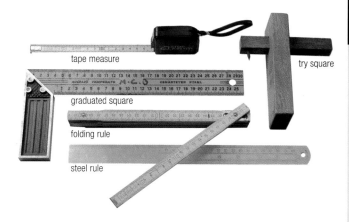

tape measure

try square

graduated square

folding rule

steel rule

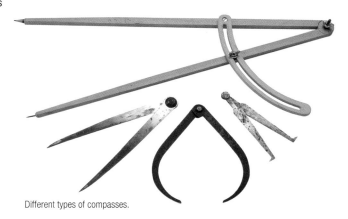

Different types of compasses.

TOOLS FOR IMMOBILIZING

When a task requires that the piece being worked remain immobile, such tools as **bench vises**, **presses**, and **pincers** are used, depending on the size of the piece and the degree of immobility required.

This type of pincer is used for pulling out nails and staples and for cutting wire.

OTHER COMMON TOOLS

Some of the most common tools that have not yet been mentioned are a **screwdriver**, which is used for tightening and loosening screws with a slotted head; wrenches that perform the same function with **nuts**; and **spirit levels**, which indicate whether or not a surface is completely level.

Three types of screw: dome head, flat head (both of which are installed using a screwdriver), and a hex screw (installed using a wrench).

Screwdrivers are used for turning screws.

MACHINE TOOLS

The term **machine** is used to designate any instrument, whether simple or complex, that is capable of performing one or more of the following actions: **modifying** the intensity or direction of **forces**; modifying or **changing the shape** of an object, or **transforming** one type of energy into another. **Machine tools** are tools that **modify** or **shape** various **materials**; however, since this definition could produce confusion by relating also to hand tools, we must add that machine tools perform their work by means of a **motor**, either internal combustion or electric, that drives them.

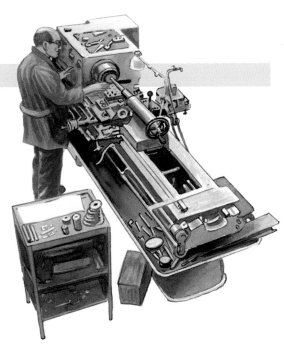

OVERVIEW OF A MACHINE TOOL

In any machine tool, in addition to the motor, there are three different types of components that can be identified: a **receiver**, a **transmission** mechanism, and a **tool** or **operator**. The receiver is commonly a simple mechanism, such as a pulley or a pinion, upon which the **motor** acts to transfer its movement. The **tool** or **operator** transforms the movement provided by the receiver and performs the work for which the machine was designed, such as shaping, drilling, or cutting; the **transmission** mechanism connects the receiver to the operator, providing the type of movement required for the work. Transmission mechanisms can be solid, such as **connecting rods**, **gear assemblies**, and **chains**, or **fluid**, such as **water**, **oil**, or **compressed air**.

An industrial lathe can be considered a representative of machine tools.

TYPES OF MACHINE TOOLS

In some cases a machine tool is really a **complement to**, and at other times a **substitute for**, the team made up of a **human** and a **hand tool**. As a result, there are as many types of machine tools as there are activities performed using hand tools. Still, some of them are used only in **industry**, whereas in comparable **crafts** or **domestic** activities the equivalent **hand tools** are still used. There are machine tools in such diverse activities as **construction**, **metallurgy**, **textile industries**, **cabinetmaking**, and **agriculture**, but we will focus on just a few particularly significant ones.

DOMESTIC APPLICANCES

There are very affordable **domestic** versions of some of these tools that function with a small **electric motor**.

Modern looms turn out fabric at high speed and with great precision.

MACHINES FOR THE LUMBER INDUSTRY

The machines used for felling trees vary according to the type of terrain and trees. In easily accessible terrain, **mechanical saws** built into a **motor vehicle** are used. In areas that are hard to get to, hand-held sabers or chain saws are used; they are powered by **electric**, **diesel**, or **gasoline** engines.

CARPENTRY AND CABINETMAKING

Wood is a material that has a tremendous variety of uses, and as a result the machines used in working it also vary greatly. In removing wood to create a **shape**, **lathes**, **power planes**, and **routers** are used; in order to assemble or **join pieces together**, **joiners**, **tenon cutters**, **punches**, **clamps**, and **glue dispensers** are used; and finally, **smoothing and polishing tools** are used in the finishing process.

A portable belt sander makes it possible to obtain a very fine finish on a wood surface.

SAWMILL MACHINERY

In a **sawmill** large logs are cut perpendicularly using **rip saws** that cut as they move along. **Radial saws** are used for cutting large, thin panels; long planks are cut using **longitudinal saws**. Usually, during these operations, the saw stays in one place and the log moves. The finest work is done using machines such as **thickness planers** that remove surface irregularities.

A band saw makes fast, precise cuts in wood.

FARM MACHINERY

In developed countries it is rare to see farmers working with hand tools or with the help of animals; it is more common to see them driving a tractor pulling some type of mechanism that varies according to the time of year or the type of crop being cultivated: This may involve a large **rake** if a new field is being prepared for sowing, or a **mechanical plow** if it is being prepared for planting. At other times farmers will use **planters** and **harvesters** that will look different based on the type of crop for which they are designed.

A tractor is a versatile agricultural tool that can perform many tasks.

Modern harvesters perform several functions at once: cutting and threshing the fresh grain, separating the grain from the chaff, bagging the grain, and baling the straw.

METALWORKING MACHINES

Machinery for metalworking can be divided into three classes according to how it functions: machines that produce **shavings or chips**, machines that change the **shape** of the metal, and machines that **physically transform** metal. In the first group there are two types of machine: ones in which the cutting **tool** remains stationary while the **part** being worked on **turns**, as with a lathe; and **drills**, **milling machines**, **grinders**, **sharpeners**, and **cutters**, in which it is the **tool** itself that rotates. The second group includes **presses** and all machines that use pressure to **fold**, **bend**, **pierce**, **cut**, or **inlay** metals. The most representative tools of the last group are **welding torches**, which melt part of the metal, as well as **cutting torches**, including those that **vaporize** the metal by means of a **laser**.

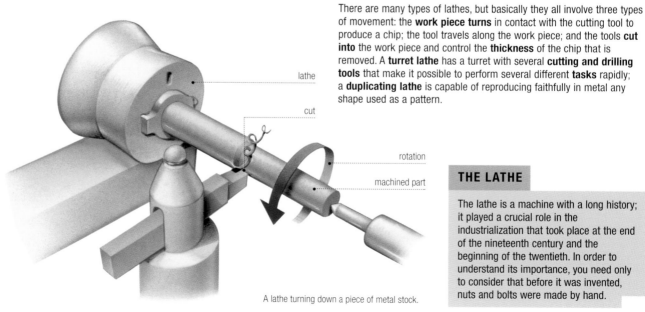

Welding joins metal parts strongly and precisely.

A single machine tool can drill several holes in a piece, turn it, reduce it, and polish it in preparation for the next stage.

PEOPLE AND MACHINES

No matter how precise and automated a machine may be, there is always a person who controls how it works.

HOW A LATHE WORKS

There are many types of lathes, but basically they all involve three types of movement: the **work piece turns** in contact with the cutting tool to produce a chip; the tool travels along the work piece; and the tools **cut into** the work piece and control the **thickness** of the chip that is removed. A **turret lathe** has a turret with several **cutting and drilling tools** that make it possible to perform several different **tasks** rapidly; a **duplicating lathe** is capable of reproducing faithfully in metal any shape used as a pattern.

lathe

cut

rotation

machined part

THE LATHE

The lathe is a machine with a long history; it played a crucial role in the industrialization that took place at the end of the nineteenth century and the beginning of the twentieth. In order to understand its importance, you need only to consider that before it was invented, nuts and bolts were made by hand.

A lathe turning down a piece of metal stock.

SMART MACHINES

Before computers became popular, there were some machines such as lathes that were able to perform certain simple movements **automatically**; however, true **automation** appeared simultaneously with **microchips**. Little by little, machines have been turning into true **robots** that perform their tasks automatically, and people become involved only in **programming** their operation, **overseeing** their correct functioning, and performing certain **repairs**. There are more and more factories where **computers** control the entire manufacturing process, immediately **readjusting** the machines as required, **notifying** the operators when the tools become worn, and interrupting the process when **changes** or **repairs** need to be made.

Robots perform their duties precisely and tirelessly; they are perfect for working in areas that present dangers to humans.

SPECIALIZED MACHINES

Technological advances in the last 30 years have created new industries that require new types of machine tools. For example, producing miniaturized electronic materials such as **microprocessors** requires highly **specialized** and **very precise** machinery. That is true also in other industries such as **aerospace**, where the production of sophisticated observation and communications satellites continues to gain importance.

ASSEMBLY LINE MACHINERY

A significant amount of industrial work is done by linking operations in **assembly lines** comprising a series of **assembly line machines** that automatically work on parts that move along them. When a part stops in front of a machine, it is worked on with one or more tools, and then is automatically moved toward another machine.

Installing machine tools has freed many people from the most laborious duties.

A REPRESENTATIVE ASSEMBLY LINE

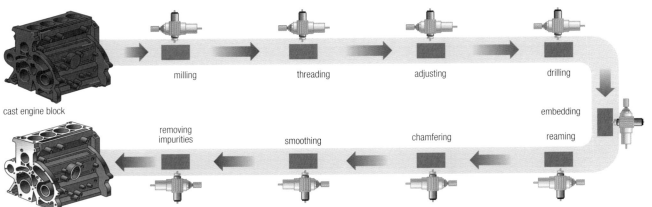

cast engine block

milling

threading

adjusting

drilling

embedding

reaming

machined block

removing impurities

smoothing

chamfering

An engine block automatically undergoes several operations by different machines.

THE STEAM ENGINE

Anyone who has seen a covered container of boiling water knows the **force** contained in **water vapor**, since it is capable of lifting the top of the container. This fact was not unknown in ancient times, and as early as the first years of the Christian era, Heron of Alexandria invented several devices that took advantage of that power. Nevertheless, a true steam engine was not developed until the **seventeenth century**, since people first had to understand certain **laws of physics** that took a long time to discover.

STEAM ENGINE OR VACUUM MACHINE?

Although it looks like a novelty item, what the physicists and engineers who perfected the steam engine really were looking for was a suction machine that could create a vacuum in one of its compartments. In the mid-seventeenth century several physicists had discovered the exceptional **force** that is produced when a great **difference of pressure** is created by **extracting** a large portion of the air contained in a hermetically sealed container. At that time the mining industry needed a machine with those characteristics, which would be capable of removing water continuously from increasingly deep mine shafts.

THE MAYOR OF MAGDEBURG

Otto von Guericke, a physicist and the mayor of this German city, called his fellow citizens to the town square to show them his latest invention. He placed two hollow metal hemispheres together with no fastening and used a hand pump connected to a valve installed on one of the hemispheres to extract the air contained inside. When he judged that the hemispheres were sufficiently empty, he had team of horses harnessed to each of the handles on opposite ends of the sphere. As long as the valve was closed, the horses could not pull the two parts of the sphere apart, but when the valve was opened the two halves came apart by themselves.

Otto von Guericke demonstrating the power of a vacuum.

EVOLUTION OF THE STEAM ENGINE

Some of the engineers who were interested in mine flooding got the idea that they could use steam power to create the vacuum necessary to pump out the water. The first elementary attempts consisted of a **boiler** closed off by a **piston**. When the water **boiled**, the steam moved the piston up to a stop, then the boiler was cooled suddenly, the **steam** condensed into a few drops of water, and a **partial vacuum** was created that applied suction to the piston; this movement was used to pump the water from the mine shafts. The invention was not very efficient, and it involved a certain amount of danger despite the improvements introduced around 1700 by the English blacksmith Thomas Newcomen.

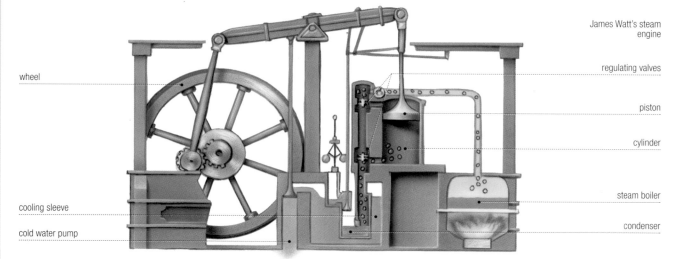

James Watt's steam engine

wheel

regulating valves

piston

cylinder

steam boiler

cooling sleeve

condenser

cold water pump

WATT'S ENGINE

Around 1770 the University of Edinburgh awarded precision mechanic James Watt the task of studying and perfecting Newcomen's invention. Watt observed that a large part of the problems were created when the **boiler** was cooled rapidly, so he had the steam travel to a new chamber, which he called the **condenser**, some distance from the first one, where the steam cooled and created the vacuum used to move a piston. That way the boiler remained hot and was three times more powerful than Newcomen's invention, as well as minimizing the danger of explosion. In 1782 Watt's engine was perfected to the point that it entered into general use.

Because of its weight and size, the steam engine was installed in a boat (1787) before it was adapted to a land vehicle (1814).

FIRST APPLICATIONS OF STEAM ENGINES

Following its success in removing water from mines, Watt's engine was put to a use that made it very popular. Up to that time, part of the energy that the textile industry needed was obtained from **water wheels**, which required locating factories next to rapid river currents; however, **Watt's engine** made it possible to locate factories in **central locations** and where there was plenty of available labor. Other manufacturers became aware of the engine's possibilities, and its widespread use signaled the start of the **industrial revolution**. In 1787 the first steam-powered boat was constructed, and in 1814 the first locomotive used it as a source of power.

Even though steam engines are scarcely used today, the steam produced by other means is used to drive turbines in the electricity industry. The photos show thermal power plants (coal, at left, and fuel oil).

CROSS SECTION OF A STEAM LOCOMOTIVE

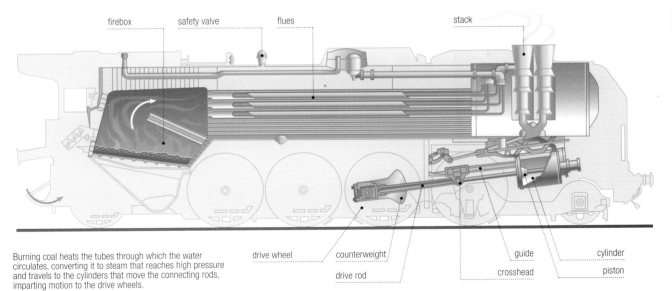

firebox · safety valve · flues · stack

drive wheel · counterweight · guide · cylinder · drive rod · crosshead · piston

Burning coal heats the tubes through which the water circulates, converting it to steam that reaches high pressure and travels to the cylinders that move the connecting rods, imparting motion to the drive wheels.

For more than a century, locomotives used the steam engine.

INTERNAL COMBUSTION ENGINES

In contrast to steam engines, where the boiler that produces the steam is separate from the pistons and the drive rods that transfer the motion, with **internal combustion engines** the fuel is burned **inside** the cylinders where the pistons that generate the movement are located, so they are far more compact than steam engines; also, even though they take up much less space, they generate more energy. The most common internal combustion engines are the ones used in automobiles, but there are other types as well.

FOUR-CYCLE ENGINES

Although fairly successful internal combustion engines had already been developed—such as the one constructed by the Frenchman Lenoir in 1860—the one built by the German technician Nikolaus Otto in 1876 became the model on which the future gasoline motors would be based. His engine consisted of a **piston** that fit perfectly inside a **cylinder** and connected to a **connecting rod** fitted to a **drive shaft** that turns on an **axis**. It also needed a **spark plug** to generate the spark, plus two **valves**, one for **fuel intake** and the other for **exhausting** the gases produced by the **explosion**.

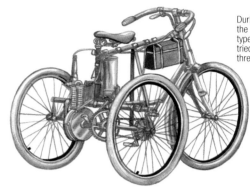

During the infancy of the automobile, many types of motors were tried in vehicles with three and four wheels.

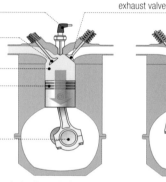 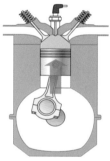 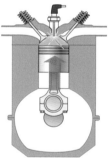 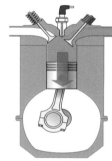

spark plug

intake valve

cylinder

piston

connecting rod

exhaust valve

The modern four-cycle engine is based on the one invented by the German Nikolaus Otto.

In the first cycle (intake), the piston descends and takes in the mixture of fuel and air through the intake valve.

In the second cycle (compression), both valves close.

In the third cycle (combustion), the spark generated by the spark plug ignites the mixture of fuel and air.

In the fourth cycle (exhaust), the valve opens and the burned gases are expelled.

Mass production of automobiles in the early 1900s started the most important industrial period of the twentieth century. The photo shows Buster Keaton and "Fatty" contemplating their car, which has fallen apart.

OTTO'S CYCLE

The four movements produced by the motor created by the German technician are known as **Otto's cycle**, and they are essentially the same for all gasoline engines. In the **first** or **intake cycle**, the piston is moved by the starter motor, although in the old days this movement was imparted by a hand crank; at that instant a **valve opens** to permit the entry of the mixture of fuel and air. In the **second** or **compression cycle**, the piston compresses the mixture so that the spark from the plug will ignite it in the **third** or **explosion cycle**. The expansion of the gases produced by the explosion displaces the cylinder; as it moves, it **expels** the gases during the **fourth cycle**.

THE FIRST MOTORS

The first attempts to create an internal combustion engine were carried out at the beginning of the nineteenth century. The first fuel used was gunpowder, but that was quickly abandoned in favor of other fuels such as vapor of turpentine and hydrogen, although gasoline was destined to become the perfect fuel for the first truly useful internal combustion engine.

EVOLUTION OF THE INTERNAL COMBUSTION ENGINE

When Otto's motor appeared, it was seen that its power could be increased by adding another cylinder, for when one cylinder was in **recovery**, the other was performing work. Soon there were four-, six-, and eight-cylinder motors arranged in a **line** or by opposing pairs in a V-shape. In 1923 an electrical system was added to operate a **starter motor**, and since that time internal combustion motors have been improving slowly but steadily.

Modern automobiles are true technological gems in which comfort and safety are as important as performance.

CURRENT INTERNAL COMBUSTION ENGINES

Even though there are very different types of motors based on the number of cylinders and their spatial arrangement, modern internal combustion engines have a number of characteristics in common that we will attempt to summarize. The main part of the motor is comprised of a metallic **block** with the cavities for the **cylinders**. There are metal **sleeves** that line the sides of the cylinders and a cylinder head that closes them at the top. The **spark plugs** and the **intake** and **exhaust valves** are located in the cylinder head; the intake valves introduce the mixture of air and fuel, and the **exhaust valves** allow for the expulsion of gases. The block is closed at the bottom by the **crankcase**, which provides a place for storing the oil that lubricates the motor. The **pistons** move inside the cylinders and are connected by a **crankshaft** that converts the straight-line movement of the pistons into a rotational movement.

THE MAIN COMPONENTS OF AN AUTOMOBILE

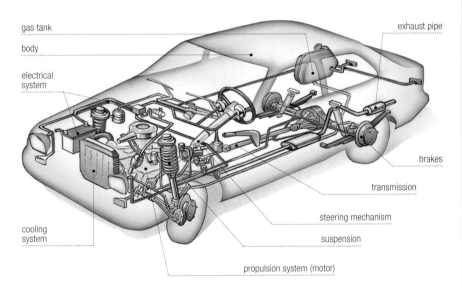

gas tank
body
electrical system
cooling system
exhaust pipe
brakes
transmission
steering mechanism
suspension
propulsion system (motor)

THE DIESEL ENGINE

In 1892 the German engineer Rudolf Diesel created an internal combustion motor that worked on **different principles** than those based on **Otto's cycle**, since it depended not on the explosion of the fuel, but rather on its **gradual combustion** as it enters the cylinder. In order to accomplish that, during the **intake** and **compression** cycles, only air is injected into the cylinder. At the point of maximum compression, the **temperature** inside the cylinder is so **high** that when the **fuel** is **injected** it is totally burned up as is it is introduced. The **precision** that this operation requires makes the **fuel injector** the most important part in its proper functioning.

BASIC DIAGRAM OF A DIESEL ENGINE

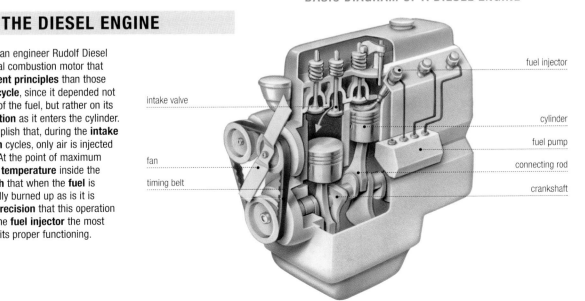

intake valve
fan
timing belt
fuel injector
cylinder
fuel pump
connecting rod
crankshaft

THE EVOLUTION OF THE DIESEL ENGINE

The need for the **compressed air** to achieve a **high temperature** before injecting the fuel required that diesel motors be very solid in order to withstand the high pressures to which they are subjected. Heavy diesel engines were not very well suited to early automobiles, and as a result they were installed primarily in **ships**, **locomotives**, **trucks**, and other types of **industrial vehicles** in which the weight of the engine was not a major factor. Another problem that the first diesel engines had to overcome was the lack of **appropriate fuel**, since gasoline was too volatile, and petroleum-based oils burned too slowly. **Diesel fuel** was introduced in the early decades of the twentieth century.

Diesel motors revolutionized heavy transport of humans and goods starting in the twentieth century, and they facilitated the introduction of new and powerful locomotives.

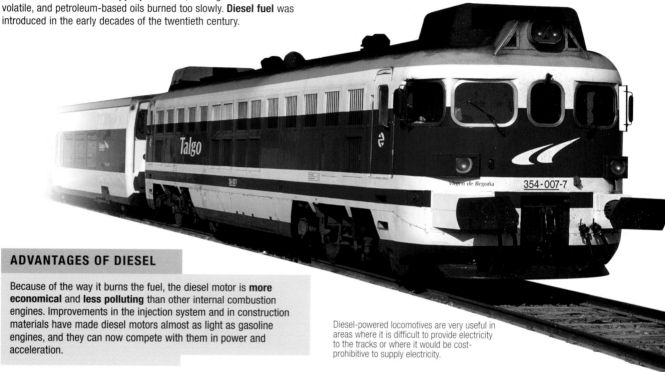

ADVANTAGES OF DIESEL

Because of the way it burns the fuel, the diesel motor is **more economical** and **less polluting** than other internal combustion engines. Improvements in the injection system and in construction materials have made diesel motors almost as light as gasoline engines, and they can now compete with them in power and acceleration.

Diesel-powered locomotives are very useful in areas where it is difficult to provide electricity to the tracks or where it would be cost-prohibitive to supply electricity.

TURBOREACTORS

Turboreactor motors are used primarily in **aviation**, although there are special versions for use in water-going vehicles. They way they operate is very simple: The motor captures **air** from the outside and compresses it inside. The compressed air **heats up** in such a way that when the fuel is injected it burns and creates expanding gases; as these are exhausted, they move the blades of a **turbine**, which, like a propeller, creates **forward propulsion**. These are mechanically simple motors that require little maintenance, but they have to be constructed with **metal alloys** that can withstand the **high pressures** and **temperatures** produced in the combustion chamber.

DIAGRAM OF A TURBOREACTOR

combustion chamber

turbine

air compressor

fuel intake

exhaust

JET PROPULSION ENGINES

These are the well-known **jet** or **reaction** engines. They are used in modern **passenger planes** for long-distance travel and for space rockets. They don't use turbines; rather, it is the **force of the reaction** to the expulsion of the gases that makes the device move, much as the air that escapes from a balloon as it deflates makes it fly. Rockets also carry their own **oxygen** in order to facilitate combustion outside our atmosphere. The **violent combustion** that is required for the expanding gases to **propel the vehicle** means that the **combustion chamber** of jet engines has to be extremely **strong**.

LIQUID FUEL ROCKET

guidance system

alcohol tank

liquid oxygen tank

alcohol and oxygen supply tubes

combustion chamber

stabilizing wings

Rockets and spaceships use powerful jet engines.

ACTION AND REACTION

It is a proven fact that forces act in **pairs** according to the principle that every force generates another force of **equal intensity** that works in **the opposite direction**. So why don't they cancel out one another? Simply because they don't act on the same object. The force that propels the charge from a shotgun when it is fired generates a **reaction** that is equal but opposite; it is applied not to the **charge**, but to the **shotgun**, which therefore **recoils**.

OTHER MOTORS

There are some motors that are not internal combustion engines, but that rather function according to totally different principles. Some of these will be described in detail farther on; here we will merely mention the **electric motor**, which converts electric energy into rotational movement, and the **compressed air motor**, which, in conjunction with electricity, uses compressed air to produce rectilinear movement in a piston.

Designers of new motors will have to be more conscious of energy efficiency and pollution reduction.

Electric motors can be found inside some appliance in practically every part of our homes: dishwashers, refrigerators, garbage disposals, coffee grinders, juicers, fans, vacuum cleaners, and so on.

ENERGY: FOSSIL FUELS

In early times, people had only their muscles to count on for performing the tasks that made it possible to survive. When people mastered fire, they significantly improved their living conditions, and with time, they used it to make new objects such as ceramics and bricks. They learned to domesticate animals and used their strength for hard and tedious labor; later on they used the strength of river currents and the wind. Until just a few hundred years ago these were the only forms of energy that people had at their disposal.

FOSSIL FUELS

The great quantity of **heat** that some substances give off when they burn is useful not only for keeping warm and for cooking, but it is also a form of **energy** that makes it possible to modify materials and make certain mechanisms function. **Wood** was the most commonly used fuel for thousands of years, but the continuous increase in population and changes in the way of life would have endangered all the forests of the earth if new fuels had not been discovered.

COAL

Coal was the first major substitute for wood once people learned how to find the large **deposits** and developed **mining techniques** that allowed its **widespread use**. Like **petroleum** and **natural gas**, coal was formed in the earth's subsoil through the accumulation and decomposition of enormous masses of **organic material** from dead vegetables and animals that were transformed by the action of fungi and bacteria over the course of millions of years. Coal deposits can be found very close to the surface as well as deep underground, and there are different techniques for extracting it based on its location.

Millions of years and certain conditions are required for a forest to transform itself into oil and gas.

COAL MINING

When coal **deposits** are very close to the surface, they are mined in **open pits** using huge, specialized machines. In some places huge **excavating wheels** are used; their toothed jaws rip up about 30 cubic yards (meters) of earth. The extracted coal falls onto a **conveyor belt** that carries it to the warehouses. Still, most coal mines are **underground**, and the main problems that they entail are **supporting the ceilings of the** **shafts** and methods of getting the coal to the surface. In the most modern operations, a **cutting machine** extracts the coal, which falls onto a **conveyor belt** as **mechanical supports** are installed behind it to hold up the roof.

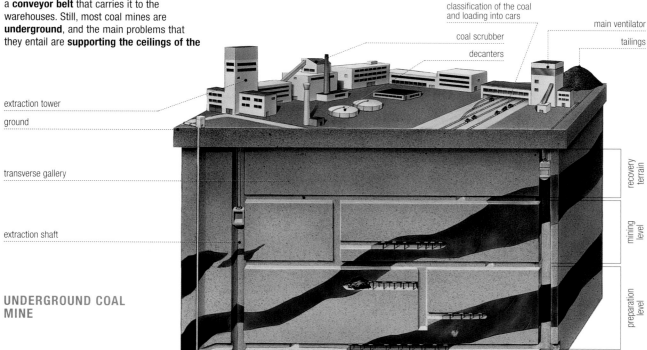

extraction tower

ground

transverse gallery

extraction shaft

UNDERGROUND COAL MINE

classification of the coal and loading into cars

coal scrubber

decanters

main ventilator

tailings

recovery terrain

mining level

preparation level

PROCESSING COAL

First the coal has to be **separated** from other rocks by dumping it into a **stream of water** strong enough to move it, since coal is light in weight, so it can be separated from heavier rocks. The part of the coal that is used for producing **coke**, an essential ingredient in steel production, is subjected to **carbonization** in large ovens at over 1,650°F (900°C). During this process, where the coal does not reach combustion, several types of gases and coal tar are given off. Further treatments for coal involve separating the **sulfur compounds** that it contains, which are a major source of pollution.

PETROLEUM

In its natural state, petroleum is a thick, sticky liquid that is found dispersed in porous underground rocks located at a depth between 450 and 25,000 feet (150 to 8,000 m). Areas that may contain petroleum, based on their **geological characteristics**, are studied by taking **samples** of air and soil and by setting off small underground explosions that are registered on **seismographs**, which produce images that are studied very carefully. When all the results point to the existence of **petroleum deposits**, **exploratory drilling** is carried out.

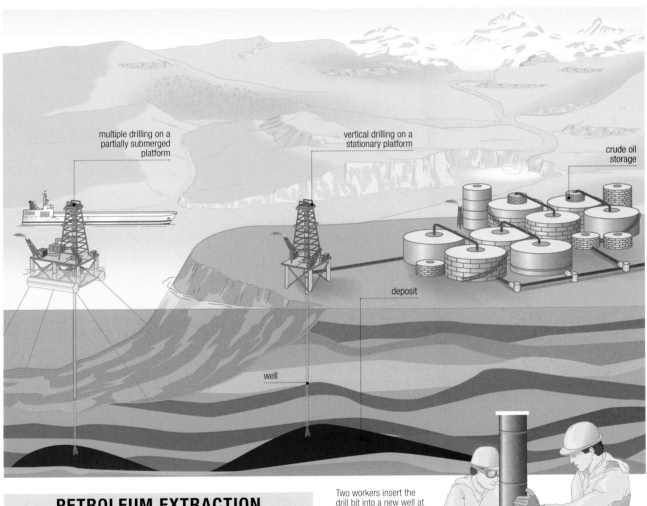

multiple drilling on a partially submerged platform

vertical drilling on a stationary platform

crude oil storage

deposit

well

PETROLEUM EXTRACTION

Drilling is carried out by means of a **drill** that turns inside a thick-walled casing that advances along with it. As the drill goes deeper, the crushed rock and mud are conveyed upward inside the casing to the mouth of the well, where the **turntable** is located; this is where the sections of the drill and the casing are fit together as the hole deepens. In order to facilitate this task, a **drill tower** is constructed that makes it possible to assemble the sections very precisely. The arrangement is completed by the **motor** that makes the drill spin, plus the **pumps** and the casings required for extracting the mud. When the deposit is reached, the well is **covered** with cement, leaving a hole about 3 inches (8 cm) in diameter through which a new tube is inserted to **isolate** the productive part of the well. To make the oil start gushing, an **explosive charge** is inserted and detonated from the surface.

Two workers insert the drill bit into a new well at the drilling site.

REFINING PETROLEUM

Crude oil is a mixture of **hydrocarbons** that need to be separated to create useful products such as **gasoline**, **diesel fuel**, and **lubricating oils**. This task is performed in large industrial complexes known as **refineries**, which sometimes are located near extraction areas, but generally are very far away from them. The main task that is carried out in the refineries is **fractional distillation**, where the fact that **different components of petroleum boil** at different temperatures is exploited. The crude oil is pumped to a tower about 140 feet (45 m) high at the base of which is an oven heated to over 650°F (350°C). The gasified oil goes up in the tower, and as it cools, its different components are **deposited** in some 40 shelves located along the tower. The heaviest components are deposited on the lower shelves, and the most volatile ones such as gasoline are deposited on the highest ones.

Processing petroleum and its derivatives is carried out in refineries.

Some of the items that can be made from petroleum.

A VALUABLE RAW MATERIAL

Since petroleum derivatives are used so extensively as fuel it is easy to overlook their use as **raw materials** in producing such important products as **plastics** and **synthetic rubber**. And that is not the end of the uses for petroleum; it is also used in many other processes in the **organic chemical industry**, as in the production of **synthetic fibers** and the **pharmaceuticals industry**.

ADVANCED PROCESSING PROCEDURES

Some heavy components of oil such as **diesel fuel** can be subjected to great **pressure and temperatures** over 930°F (500°C) to produce **gasolines** in a process known as **thermal cracking**. A similar effect is obtained through **catalytic cracking**, which is in far greater use today and doesn't require such high temperatures or pressures. This requires putting the petroleum into contact with a substance known as a **catalyst**, which improves the procedure but is not used up in the process.

Petroleum can be distributed by means of pipelines or tank trucks.

NATURAL GAS

Natural gas is a fossil fuel that was formed much like petroleum; it is extracted by processes similar to those used with crude oil. In the natural state it is a mixture of gases, primarily **methane**, but it also includes **undesirable products** that must be removed before distributing it. It is used mainly as a **fuel**, but it is also an important **raw material** in some **petrochemical** processes.

domestic distribution network

liquefaction plant

methane ship

gas pipeline

deposit

regasification plant

methane ship

The natural gas that we consume travels thousands of miles (kilometers) to reach our homes. This illustration shows the gas distribution network.

Natural gas is stored in liquid form in large storage facilities before being distributed.

TRANSPORTATION AND STORAGE

Once the natural gas is purified, it can be transported through **gaslines** or **tanker ships**. Since natural gas can withstand great pressure without turning to liquid, gaslines thousands of miles (kilometers) long are constructed; these require **recompression** stations about every 60 miles (100 km). Using tanker ships as a means of transportation requires prior conversion of the gas into **liquid form**. This is carried out in **liquefaction plants** where the temperature of the gas is lowered to −256°F (−160°C). The resulting liquid takes up just 6 percent of the volume that the gas occupied at ambient temperature. The tankers in which it is transported have to maintain this low temperature until it is unloaded at regasification plants, where the gas is channeled to the areas where it is consumed.

Even though new deposits have been discovered in recent years and old ones are being exploited more efficiently, fossil fuels are a finite resource.

Introduction

Machines and Tools

The Steam Engine

Internal Combustion Engines

Energy

Electricity Production

Electric Motors

Mining

Metallurgy

The Chemical Industry

Public Works

Transportation Vehicles

Imaging

Electronics

Computer Science

Robotics

Index

ELECTRICITY PRODUCTION AND DISTRIBUTION

Lightning is the most spectacular manifestation of electricity in nature, but the effects of natural static electricity also become evident when we put our hand close to some metal object on very dry days and we experience a mild electric shock. Despite this constant presence of electricity, humans were not able to study it until the eighteenth century, and it was only in the following century that the first instrument capable of producing an electric current was devised.

THE NATURE OF ELECTRICITY

Electricity is a form of energy that the most **elementary components of matter** possess. **Electrons** are packets of pure electric energy that are commonly designated as **negative**. **Protons**, on the other hand, have as much electrical energy as electrons, but of a type that is designated as **positive**. The effects of the two types of electricity **neutralize** one another, and as a result, **atoms**, which contain as many electrons as protons, are electrically neutral. In order for matter to demonstrate electrical properties it must be **ionized**; in other words, one of its atoms has to have **lost** or **gained** one or more electrons.

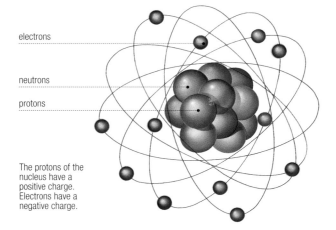

electrons

neutrons

protons

The protons of the nucleus have a positive charge. Electrons have a negative charge.

THE FIRST ELECTRICAL GENERATORS

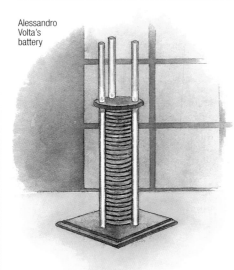

Alessandro Volta's battery

In order to produce a continuous stream of electricity, some type of force must be applied so that **electrons** that are not tightly bonded to a **conductor** follow a **certain direction** in an orderly fashion. The first way that was discovered to create this effect involved some **chemical reactions** that were capable of **ionizing** certain substances. The Italian physicist Alessandro Volta stacked disks of **copper** and **zinc** separated by cardboard disks moistened in salt water, and connected two copper wires to the ends of this stack; he thus created the first **battery** capable of generating an **electrical current**.

The same principle was applied early in the nineteenth century to other batteries, which were the forerunners of both the **dry cell batteries** used in countless domestic devices and **automobile batteries**.

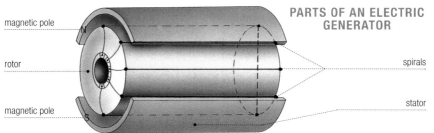

PARTS OF AN ELECTRIC GENERATOR

magnetic pole

rotor

magnetic pole

spirals

stator

ELECTRICAL CONDUCTORS AND INSULATORS

The ability to give up or take on electrons is what gives different materials their properties as **conductors** or **insulators**. Metals such as **copper**, **silver**, and **aluminum** are good electrical conductors because they have electrons that are bonded weakly, and easily leave the atoms. **Glass**, **wood**, **porcelain**, and **plastics** are insulating materials because their electrons are very stable.

The objects on the left (plastic, glass, and ceramic) are not good conductors of electricity; those on the right (metals) are.

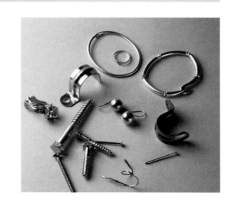

Batteries are useful for producing **small quantities** of electricity, but they are extremely **expensive** when **significant amounts** of electricity need to be produced. Using the current produced by batteries made it possible for physicists to observe some phenomena that were produced in conductor wires, such as the fact that two nearby, **parallel cables** in which **current** was running in the **same direction** were **attracted to one another**, whereas when the current was running in **opposing directions** the cables **repelled one another**. Also, when a compass was placed near a cable in which electricity was running, the needle changed direction. Inspired by these observations, the French physicist André-Marie Ampere demonstrated that a cylindrical coil of **conductor cable** behaved like a **magnet** when **electrical current** ran through it.

ELECTRICITY AND MAGNETISM

Opposite poles (left) attract, whereas similar poles (right) repel one another.

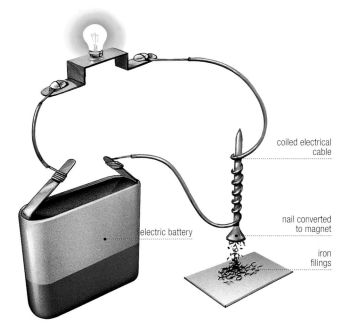

coiled electrical cable

electric battery

nail converted to magnet

iron filings

FARADAY'S GENERATOR

The English scientist Michael Faraday had the idea that if electricity was capable of producing magnetic effects, perhaps magnetism could produce electricity, so in 1831 he attempted to **induce** an electric current into a coil placed near an **electromagnet**. The experiment was a failure, but it allowed him to discover that it was the **movement** of a **conductor** in the **vicinity of a magnet** that produced the **electrical current**. In order to be sure, he created a device in which a **copper disk rotated** continuously between the poles of a magnet in the shape of a horseshoe. By connecting a voltmeter to the apparatus he demonstrated that while the disk was turning it produced electrical current. He had invented the first **electrical generator** that was not based on chemical principles.

It is easy to make an electromagnet and test its properties. By setting up a circuit like the one in the illustration, the iron filings are attracted by the nail, since the electricity of the battery has converted it to a magnet.

INDUSTRIAL PRODUCTION OF ELECTRICITY

Based on Faraday's invention, physicists and engineers realized that if they **connected** a more highly developed electrical **generator** to an inexpensive **energy source** that caused it to move continuously, they could produce **abundant and cheap electricity**. Soon they constructed small electrical power generating plants that used both **running water** and the motive force of **steam engines**. They quickly encountered the problem that there are **two different ways** of producing electricity: one produced current flowing in a single direction, therefore referred to as **continuous**; and the other changed the direction of the current several times a second, so it was referred to as **alternating**. This discrepancy complicated the production of devices capable of using electricity to function.

PARTS OF FARADAY'S DYNAMO

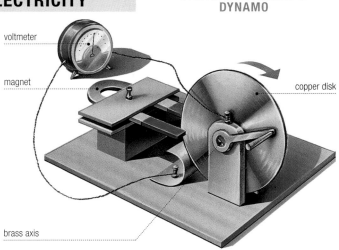

voltmeter

magnet

copper disk

brass axis

HYDROELECTRIC POWER PLANTS

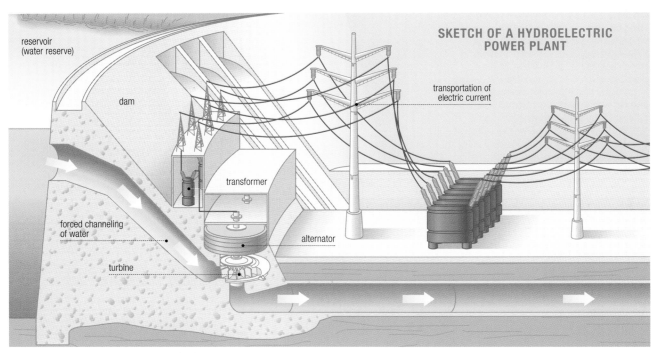

SKETCH OF A HYDROELECTRIC POWER PLANT

reservoir (water reserve)

dam

transportation of electric current

transformer

forced channeling of water

turbine

alternator

PARTS OF PELTON'S HYDRAULIC TURBINE

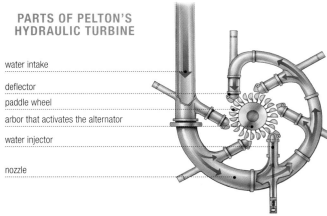

water intake

deflector

paddle wheel

arbor that activates the alternator

water injector

nozzle

Once people solved the problem of the type of current to produce, one of the cheapest ways to produce it was to take advantage of running water to power the generators; however, the **sudden variations** in the **rate of flow** demonstrated the need to create **millponds** or **reservoirs** that could be regulated to use their energy in a more **constant** manner throughout the year. In order to take better advantage of the power of the differences in water level, **hydraulic turbines** were designed for the conditions of each reservoir.

PELTON'S WHEEL

This involves an **impulsion** turbine designed to take advantage of water that is taken in under great **pressure**. The water, guided by piping, comes out through smaller orifices and strikes the paddles of the turbine **rotor**, thereby causing it to turn at high speed. This is the type of impulsion turbine most commonly used in hydroelectric power plants.

THERMOELECTRIC POWER PLANTS

In traditional thermoelectric power plants fuels such as oil, coal, and natural gas are burned to produce the **steam** necessary to move the **turbines** that make the **electrical generators** work. The **long energy transformation chain** that begins with the fuel and ends in electricity means that a fair amount of the initial energy is **lost** during the changes, making available about 50 percent of the initial energy. These power plants are also major producers of **air pollution**, since they release into the atmosphere great quantities of **carbon dioxide**, which is the major contributor to the gradual **warming** of the planet.

HOW A THERMAL POWER PLANT WORKS

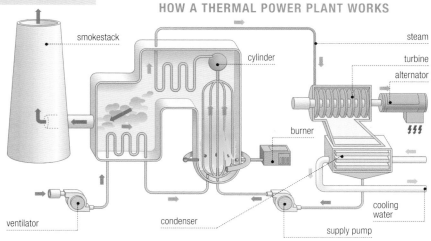

smokestack

cylinder

steam

turbine

alternator

burner

condenser

cooling water

supply pump

ventilator

THERMONUCLEAR POWER PLANTS

Thermonuclear power plants use a type of energy that was totally unknown until well into the twentieth century. **Nuclear fission** is produced when a **neutron**, which is a component of an atom's nucleus, collides violently with another atomic nucleus and **expels** part of it. This division of the nucleus releases a tremendous amount of **energy** and causes other neutrons to replicate the situation and create a **chain reaction**. This reaction is **controlled** within the **reactors** of the thermonuclear power plants so that it proceeds at the **appropriate speed** for converting the heat given off in reaction to the steam that moves the turbines and make the generators work.

SKETCH OF A NUCLEAR REACTOR AND STEAM GENERATOR

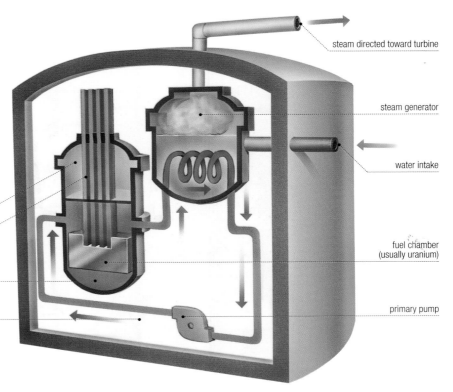

steam directed toward turbine

steam generator

water intake

reactor

control rods

fuel chamber
(usually uranium)

vat

pressurized water

primary pump

ADVANTAGES AND DISADVANTAGES OF NUCLEAR FISSION

The power plants that use nuclear fission to produce electric energy consume small quantities of **enriched uranium** as fuel, and since **a pound** (half kilo) of this material has the energy equivalent of **1,500 tons** of coal, the saving in fossil fuels is enormous, and it avoids releasing millions of tons of carbon dioxide into the atmosphere. Just the same, the solid residues that are produced, even though they are small in quantity, are **radioactive**, and they retain their power of contamination for hundreds or thousands of years. Some of these residues are **recycled**, but others are simply placed into storage. The second problem with thermonuclear reactors involves **security**, since serious accidents entail catastrophic and long-lasting consequences.

WIND ENERGY

Using **wind** to produce electricity has come a long way in recent decades. Modern **wind generators** are situated in strategic locations to create **wind farms** that are beginning to become common in Europe and other places. These modern windmills consist of a very high tower with a wind generator at the top that is comprised of three major parts: the **blades**, of which there are normally **three**; the **directional mechanism**, which automatically puts the blades into the best position with respect to the wind; and the **electric generator**, which functions through the movement of the blades.

Wind-generated electricity is produced very cleanly and without consuming nonrenewable resources; however, in some cases people criticize the location of the wind farms in areas of ecological or scenic interest.

SOLAR ENERGY

The energy from solar light can be transformed into electrical energy through two processes: using **solar ovens** or through **photovoltaic cells**. Enormous **parabolic mirrors** are used in solar ovens to concentrate the light in the center of the oven where it generates the **steam** that, as in other electric power plants, provides the motive force for the generators. The mirrors are controlled by a **computerized** system that moves them according to the position of the sun. This type of power plant can be constructed only in places that get many hours of sunlight throughout the year and where there is plenty of space for siting them. Photovoltaic cells don't require as much space because they produce the electricity directly by means of the **photoelectric effect** presented by some metals, such as **cesium**; it involves the sun's capacity to mobilize certain surface electrons in their atoms.

Solar panels are increasingly being used in building construction, especially where the buildings are located far from power distribution lines.

The Odeillo solar oven (France). Hundreds of mirrors concentrate the sun's rays onto a parabolic mirror, which in turn concentrates them onto the oven.

The sun gives off five million tons of its mass as radiation, and even though just one ten-millionth of that energy reaches the earth, today it is a promising source of energy.

PHOTOVOLTAIC PANELS: AN URBAN SOLUTION

In cities that enjoy many hours of sunlight every year, photovoltaic panels for producing electricity are cropping up on the terraces of newly constructed buildings; other solar panels are used for heating water. Since electricity can't be stored, these panels have to be connected to the general electric grid, which leads to certain bureaucratic hesitancy on the part of the companies that provide electricity. If these problems can be overcome, solar panels will become a significant part of the urban landscape.

PHOTOVOLTAIC CELL MADE WITH TWO TYPES OF SILICA

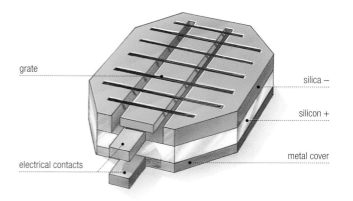

grate

silica −

silicon +

metal cover

electrical contacts

ELECTRICITY DISTRIBUTION

When electricity came into general usage it was cheaper to transport **alternating** current, so **household appliances** were manufactured for this type of current. Some decades later, methods for transporting **direct** current improved significantly, so some of the electricity that was distributed had to be **converted** from direct to alternating before it reached homes.

An amazing infrastructure is required for a light bulb to shine in the home.

High-tension tower. The electricity that reaches industry and homes flows through its copper wires.

VOLTAGE REDUCTION

In the initial stages, electricity is generated and transported at about 500,000 volts; this needs to be reduced to 110 volts in America and 220 volts in Europe, since these are the voltages for which household appliances are designed. This reduction is accomplished in **stages** in different **reduction stations** located along the grid. The electric current is transported by means of **conducting cables** made from an alloy of **copper** and **aluminum** that are usually held up by towers, although today, underground cables are becoming more common. The support points where the electrical cables contact the towers and posts are made of **glass** or **porcelain** disks that act as insulators.

Imagine for a moment what the consequences would be if we didn't have electrical energy for a period of several weeks.

About 60 percent of the electricity that our cities and industries consume is generated in thermoelectric power plants.

ELECTRIC MOTORS

An electric motor is a machine that **transforms electrical** energy into mechanical energy, in other words, into **movement**. Electric motors are an indispensable part of such common household appliances as vacuum cleaners, washing machines, and refrigerators, but in addition, they are industrial machines that produce motive force more **cleanly** and **quietly** than internal combustion motors. Electric motors perform a function opposite that of generators, but there is a marked similarity between them.

ELECTRIC MOTOR BASICS

The American physicist and inventor Joseph Henry thought that if the **movement** of a copper disk in the **magnetic field** of a **magnet** produced **electric current**, as Faraday's experiment had shown, it might also do the opposite; in other words, it might be possible to use **electricity** to produce **movement**. In order to demonstrate that, he constructed an electric motor that consisted of **two acid batteries** that produced the electricity necessary to cause two horizontal bars to move in an alternating fashion.

DIAGRAM OF JOSEPH HENRY'S ELECTRIC MOTOR

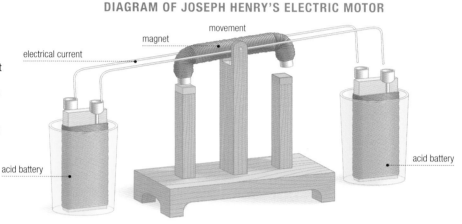

movement
magnet
electrical current
acid battery
acid battery

HOW AN ALTERNATING CURRENT ELECTRIC MOTOR WORKS

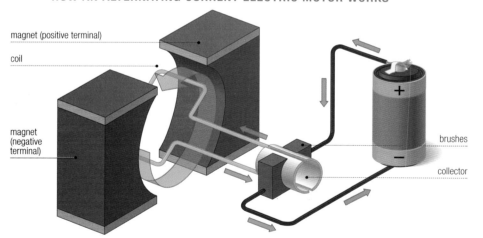

magnet (positive terminal)
coil
magnet (negative terminal)
brushes
collector

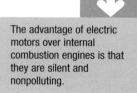

The advantage of electric motors over internal combustion engines is that they are silent and nonpolluting.

DIRECT CURRENT MOTORS

This family of electric motors includes the starter motors of automobiles and the motors in moving toys and models. Here is how they work: **external current** reaches the motor through two graphite **brushes** located on the sides of a **copper switch** where the coils of the same material originate and through which the current circulates. The coils are inside the magnetic field of a **magnet**, and when the current enters through one of the brushes, it passes through the switch and goes through half the coil. This produces a **force** that causes the switch and the coil to pivot 180 degrees, but in order for the turning to continue in the same direction, the switch **changes the direction** of the current.

Electric motors can be as large as the ones that power real locomotives or as small as a model train.

INDUCTION MOTORS

Some motors that are used in industrial machinery, and especially in electric household appliances, belong to the family of induction electric motors. These are composed of a stationary part, the **stator**, and another one that turns, the **rotor**. Both of them are lined with a **spiral winding** of electrical wire. The external current that flows through the winding of the stator **induces** current in the rotor. A magnetic field is generated in the narrow space that separates the rotor from the stator; this causes the **rotational movement** in the rotor. Since these motors use **alternating current**, the rotational speed depends as much on the **frequency** at which the electrical current is supplied—normally 50 or 60 hertz—as well as the number of spirals that make up the winding.

THREE-PHASE MOTORS

Most industrial motors are **three-phase**, for this is the type of electricity most commonly supplied to industry. The principle by which they work is the same as with other induction motors, but since three-phase current is delivered through **three** wires, instead of the **two** through which it is supplied to homes, the windings for the rotor and the stator are made accordingly.

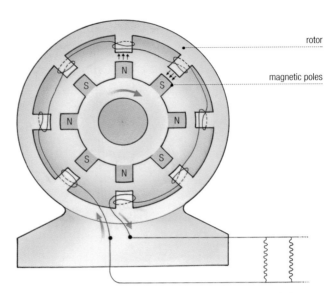

rotor

magnetic poles

In the monophase alternator, the rotor is comprised of a spiral with two conducing wires.

WAYS OF CONNECTING THREE-PHASE CURRENT

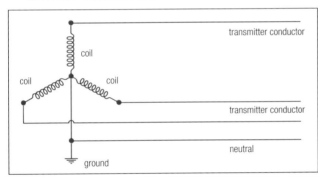

transmitter conductor

coil

coil   coil

transmitter conductor

neutral

ground

coil   coil

coil

transmitter conductor

transmitter conductor

transmitter conductor

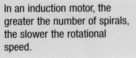

In an induction motor, the greater the number of spirals, the slower the rotational speed.

THREE-PHASE CURRENT

This is the most efficient way to distribute electricity and the most widely used throughout the world. **Monophase** conduction, which is limited to just a few domestic uses, requires **two** conducting wires, whereas using **three** conductors makes it possible to create **three-phase** conduction, which would require **six** wires with the other system. As the diagram shows, there are two way to configure the connections: in a **Y**-shape and a **delta**-shape.

Transformers are capable of converting monophase current to biphase and increasing or reducing its voltage.

MINING

Materials that are easy to access exist in limited supply, and their uses are very limited. With the exception of **copper**, which was abundant on the surface in certain areas of the planet, other metals in **mineral** form generally are found at considerable depth underground. As humans learned to identify these minerals and discovered procedures for purifying the metals they contain, extraction methods were also developed that gave rise to the mining industry.

METALLIFEROUS DEPOSITS

The areas where **mineral deposits** are located from which **metals** can be **extracted** are referred to as **metalliferous deposits**. They come in different shapes and sizes, but the most important piece of information used in deciding whether or not to proceed to mining it is the **mineral concentration**; in other words, the quantity of mineral contained in each cubic yard (meter) of earth removed. The part of the deposit where the mineral is located is called the **vein** or the **seam**; it commonly is long and narrow in shape, and there are often cracks in the rocks that have filled with the mineral.

pit mine

main shaft

elevator

elevator

vein

access gallery

access gallery

lifting station

irregular mineral deposit

parallel gallery

future mining site

drain

TYPES OF VEINS

The shape and arrangement of mineral veins are the result of **geological** activities, **volcanic eruptions**, **fractures** in the earth's crust, leaks of **magma** and **gases**, and **earthquakes**. Vertical veins are termed **chimneys**; they were formed mainly by volcanic action. **Lenticular** deposits are formed at the confluence of a fold in the terrain with a **fracture** of the earth's crust. Still, the majority of veins are irregular in shape because they have undergone many changes since the time of their formation.

This cross section of a volcano show the formation of successive layers.

MINING OPERATIONS

When the mineral is easily accessible from the surface, it is obtained by **terrace mining**. These terraces descend from the edge of the mining field by way of **ramps** that allow the movement of vehicles and personnel. The mineral is obtained by combining **explosives** and **excavating** machinery, and it is carried to an upper level by way of **conveyor belts**, **carts**, or **dump trucks**.

View of a pit mine in Ukraine.

UNDERGROUND MINING

Every underground mineral deposit requires its own geological study, since mining it depends on a variety of factors such as its **shape**, the **type of rock** involved, and the way it is arranged in the earth. In order to ascertain these factors, **controlled explosions** are carried out at different depths. These are analyzed using graphs from a **seismograph**; various **probes** are made for taking **samples**. Once it has been decided what type of mining to do, a **vertical shaft** or a **descending gallery** is dug, depending on the nature of the deposit, down to the deepest level where the desired mineral is located.

Geologists use complex equipment to study the characteristics of the subsoil to locate and evaluate possible deposits.

For centuries the dangerous conditions in many mines have caused serious accidents. The photo shows a tin mine in Potosí (Bolivia).

GALLERIES

From this main entry **access galleries** are drilled toward the mineral; these are called **level** when they are parallel to the mineral vein, or **transverse** when they intersect it. **Secondary galleries** are also needed in the mine for the installation of such **services** as workshops, first-aid stations, **ventilation**, and **extraction**. In some mines **water removal** systems are also very important, since in many cases the work is carried out in areas where the galleries would be continually **flooded** if the water weren't drained.

SHORING UP THE ROOF

Inside the gallery of a coal deposit.

When the roof of a gallery is strong enough, it is sufficient to install simple pillars at regular distances to reinforce it; still, the most common practice is to use **wooden structures** because they provide more security. In large deposits and in places where the galleries cross one another, the structure is reinforced using iron or steel **beams**. When the galleries are to remain open for a long time, **steel rods** are inserted into the rock and held in place with **plates** and **nuts**. This system is also used to hold large, **unstable rocks** in place.

PREPARING THE MINERAL

The material that is extracted from the mine is a mix of mineral and residue that are separated by **crushing** and **washing**; in some cases procedures involving **magnetism** must also be used.

METALLURGY

Metalliferous minerals require different treatments to extract the **metals** they contain. Some processes are common to the **metallurgy** of several metals, but others are used with just one. Metallurgy is the set of knowledge necessary for treating metals, and it is one branch of technology that has a fairly long history. Some metals such as **copper**, **iron**, and **tin** have been mined for more than **3,000 years**, and the processes for working them have evolved significantly during that time.

METAL PREPARATION

When the mineral is free of impurities it is subjected to various treatments to separate the metal it contains. These treatments can be carried out in three ways: through heating or **pyrometallurgy**, by using electricity or **electrometallurgy**; or by using water or **hydrometallurgy**. Metals such as **lead**, **copper**, and **zinc**, which form compounds with sulfur, are subjected to a current of **hot air** and then are made to react with carbon to liberate them. Others, such as **zinc**, **mercury**, and **cadmium**, are converted to **gas** at relatively low temperatures; they are heated with pulverized **coal**, and the metallic vapors are collected in a condenser.

STEEL METALLURGY

mineral
agglomerate
coke
coal
scrap
smelting
scrap
oxygen
liquid steel
tapping boiler
electric furnace
continuous emptying
ingots
smoothing
laminating
reheating
reheating
laminating
bars
wire
beams
rails
forms

SMELTING

This is the most widespread metallurgical process, and it is applied to many types of metal. It involves raising the **temperature** of the mineral until it changes to **liquid**, at which time it is easy to separate the metal from the **gangue** or the impurities. During the smelting of some metals, small quantities of other **elements** are often added to **improve** the **characteristics** required for the usage for which they are destined. This operation, known as **alloying**, can be done all at once or in several stages.

Each mineral melts at a different temperature. Usually it is heated in a container known as a crucible.

Metal just taken from the blast furnaces.

LOST WAX OR INVESTMENT CASTING

This is a very ancient method that has been perfected over time. It consists of filling a **refractory** mold with wax at ambient temperature; when the molten metal is poured into the mold, the wax melts and the metal takes its place. This yields seamless parts that contain great precision and detail.

FINISHED PRODUCTS

Molten metal can be converted to **ingots**, **tubes**, or **rods** in the process known as **continuous smelting** or **continuous solidification**, where the metal is allowed to flow freely from the **furnace** and is stretched out continuously while it is still molten. Depending on the quality of finish the pieces require, the molten metal can be poured into **molds** made of **sand** or **moist clay** for coarse pieces, or into metal or plastic molds, which provide a better finish and more precise detail.

ELECTROMETALLURGY

Some molten metals **ionize**; that is, they acquire an electric charge, so it is possible to **separate** them from the mixtures in which they are found by circulating an **electric current** through the molten metal in such a way that the **pure metal** is deposited on one of the **electrodes**. This is the means used to produce **aluminum**, **magnesium**, **zinc**, and **cadmium**. This same principle, which is known as electrolysis, is used for **plating** other metals with thin layers of **silver**, **aluminum**, **chrome**, **nickel**, and **zinc** to give them the shine characteristic of these metals.

HYDROMETALLURGY

Although the most popular process of hydrometallurgy is washing gold-bearing sand to isolate the gold nuggets, there are other metallurgical procedures of greater industrial significance that also use water; an example is **leaching**, which consists of **dissolving** a mineral in water to separate the metal that it contains through chemical or electrical processes. Some electrometallurgical processes are also carried out on watery solutions.

HOW ELECTROLYSIS WORKS.

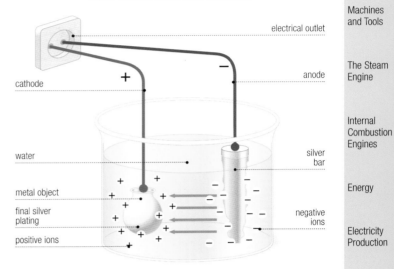

- electrical outlet
- cathode
- anode
- water
- silver bar
- metal object
- final silver plating
- positive ions
- negative ions

THE OLDEST ALLOY

AAlthough an accidental mixing of two molten metals could have happened much earlier, there is archeological evidence that around 3500 B.C. molten copper and tin were being mixed together to intentionally produce a metal alloy: **bronze**, which is much harder than either of the two metals by itself.

THE IRON AND STEEL INDUSTRY

Although such metals as **gold**, **silver**, **copper**, **tin**, and **lead** were used earlier, **iron** occupies a privileged place in metallurgy because its coming into widespread use represented a revolution in how our species lived. Despite its abundance, iron was a latecomer to metallurgy because it is practically never found in a pure state on our planet—only a few meteorites are metallic iron—and it is commonly found in minerals such as **red hematite** and **limonite**, which are iron oxides.

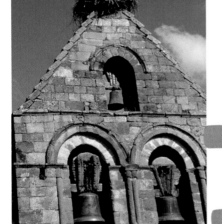

Bells are made of bronze, an alloy of 78 percent copper and 22 percent tin.

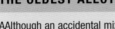

Because of its rarity, gold has been prized as a precious element ever since ancient times; it has been used for coins, jewelry, and objects that represent luxury and power.

Engraving that shows gold seekers in a California river in the middle of the nineteenth century. This activity has been immortalized in western films.

REFINING AND SMELTING IRON

The minerals that contain iron are crushed and put onto a conveyor belt where **magnets** separate the **magnetic material** that contains the **iron**. The purified mineral is converted to metallic iron in the shape of ingots in **blast furnaces**. A blast furnace is a huge cylindrical structure made of **steel** over 90 feet (30 m) high lined on the inside with **refractory bricks** that can withstand very high temperatures.

HOW BLAST FURNACES WORK

Blast furnaces are fed continuously from the top with a mix of **iron mineral**, a type of high-energy coal known as **coke**, and **tufa**, or limestone. Air is injected into the furnace at around 2,400°F (1,300°C) and a speed of 600 feet (200 m) per second; this creates a cavity in the body of heated coke, which gives off gases at over 3,600°F (2,000°C), and as they rise up the furnace they **remove the oxygen** from the iron oxides. Liquid iron drips between the shell and the few remaining bits of oxide, which are eliminated. The final product is molten iron in **ingots** that are separated from the leftover **gangue**.

Electric blast furnace.

STEEL

Steel is an **alloy** of iron that contains between .03 percent and 1.5 percent **carbon**. It is **harder**, more **rust resistant**, and **more malleable** than iron. It is smelted in ovens similar to blast furnaces, but the load that goes in with the iron is different depending on the mineral being used and the type of steel to be produced.

Columns of recently smelted iron in a British steel plant.

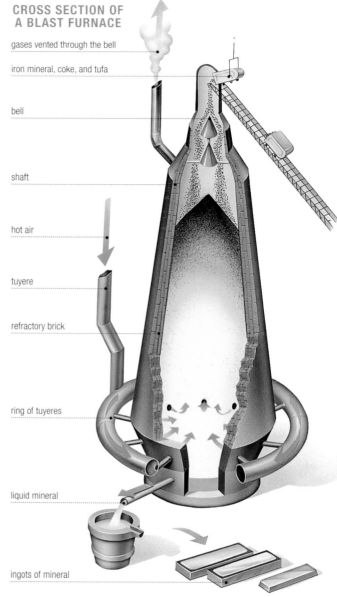

CROSS SECTION OF
A BLAST FURNACE

gases vented through the bell

iron mineral, coke, and tufa

bell

shaft

hot air

tuyere

refractory brick

ring of tuyeres

liquid mineral

ingots of mineral

ALUMINUM METALLURGY

More than 8 percent of the earth's crust is aluminum, and that makes it by far the most abundant metal in the crust. Although it is part of many common minerals, the majority of aluminum is extracted from **bauxite**, which contains about 52 percent aluminum in the form of **oxide**. The most common method for obtaining aluminum is the **Bayer process**, which starts by grinding the bauxite and mixing it with **caustic soda**. The result of the reaction is **filtered**, **clarified**, and **cooled**; then hydrated aluminum crystals are added to trigger the **crystallization** and **precipitation** of **aluminum oxide**. This is **filtered** and **heated** to 1,800°F (980°C) in rotating furnaces from which a white powder similar to sugar is extracted; this is **pure aluminum**.

USES FOR ALUMINUM

Aluminum oxide is the raw material from which metallic aluminum is obtained by **electrometallurgical procedures**. Pure aluminum is malleable and soft, but not very durable; its alloys with **copper** and **magnesium**, however, acquire **hardness** and **resistance to corrosion**, and are easy to **weld**. As a result, they are used in manufacturing **screws**, **structures**, and components for the **aeronautics industry**. The alloys with **manganese** have a broad application in **kitchen utensils**, **chemical** implements, **storage tanks**, and **architectural** elements.

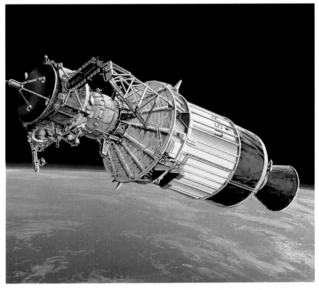

Some aluminum alloys are very strong and are used in satellite construction.

COPPER METALLURGY

Copper is obtained principally from minerals such as **cuprite**, **chalcocite**, **and chalcopyrite**, which are ground and separated by **selective flotation**. The useful part is subjected to **roasting** by means of hot air currents, and then it is **melted** to separate the **gangue**. The resulting **imperfect melt** needs to be **refined** in a subsequent melt done in **reverberation furnaces** or **electric furnaces**, where it is once again subjected to air currents so that the remaining iron and sulfur are oxidized.

Many sculptures made for display outdoors are made of copper. This sculpture is by Fernando Botero.

USES AND ALLOYS OF COPPER

Pure copper is used in manufacturing components for the **electrical industry**, but copper alloys are also important in industry. **Brass**, which is used in **turning** and the furniture industries, contains more than 55 percent copper and less than 45 percent **zinc**. **Bronze**, which is used for casting, can contain between eight and sixteen parts copper for every part of **tin**. The most common form of **cupronickel** contains 30 percent nickel; it is used for making tubing in which very hot water vapor circulates.

The cables for high-tension towers are made of an alloy of copper and aluminum, which are good conductors and very strong.

THE CHEMICAL INDUSTRY

The vast majority of objects that we encounter on a daily basis are not made of materials that can be found in nature, since they are made entirely by the chemical industry, or they have undergone major chemical transformations. The **chemical processes** are noteworthy because the **substances** that are used initially **are transformed** to create others that have entirely different properties. These **reactions** are produced in special conditions that make up **chemical technology**.

THE DIVERSITY OF THE CHEMICAL INDUSTRY

There are so many products and processes related to the chemical industry that it is difficult to classify them. From a practical standpoint, they can be divided according to the size of their installations in **heavy** industry, when they serve substantial installations with **large-scale**, **permanently installed** equipment and employing a **small number** of people; and **light** installations using a variety of equipment of **moderate** size, which often is **mobile**, and assembled on industrial sites where **many people** work. If the raw materials used are **petroleum** derivatives, we speak of **organic** or **petrochemicals**; otherwise the term is **inorganic**.

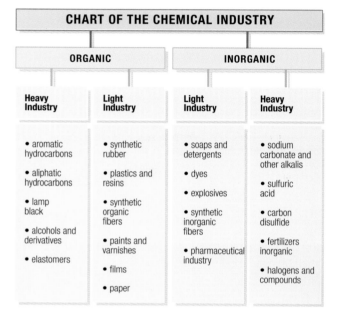

CHART OF THE CHEMICAL INDUSTRY

ORGANIC		INORGANIC	
Heavy Industry	**Light Industry**	**Light Industry**	**Heavy Industry**
• aromatic hydrocarbons	• synthetic rubber	• soaps and detergents	• sodium carbonate and other alkalis
• aliphatic hydrocarbons	• plastics and resins	• dyes	• sulfuric acid
• lamp black	• synthetic organic fibers	• explosives	• carbon disulfide
• alcohols and derivatives	• paints and varnishes	• synthetic inorganic fibers	• fertilizers inorganic
• elastomers	• films	• pharmaceutical industry	• halogens and compounds
	• paper		

Glass comes from the fusion of a complex mixture in which silica, or plain sand, predominates.

SULFURIC ACID PRODUCTION

This is produced primarily by the **contact method**, using **sulfur dioxide** obtained from the oxidation of sulfur. This gas is **purified** and oxidized again in an **oven** at a temperature between 750 and 1,100°F (400 and 600°C), where it comes into contact with a **vanadium salt**. The resulting **sulfur trioxide** is made to react with **water** to produce **sulfuric acid**.

THE HEAVY INORGANIC CHEMICAL INDUSTRY

This industry manufactures products that are used as raw materials for other industries, using inexpensive and very abundant materials such as **common salt**, **air**, **water**, **coal**, and **limestone**. The most important product from this branch of the chemical industry is **sulfuric acid**, but **sodium carbonate** and commercial **soda**, which are used in producing **glass** and **soaps**, are also important, as are **ammonia** and **nitric acid**, which are used to produce **fertilizers**, and **chlorine**, which is used in the **paper** and **textile** industries.

THE PROCESS OF OBTAINING SULFURIC ACID

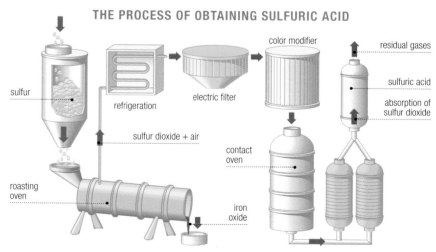

sulfur

refrigeration

color modifier

residual gases

electric filter

sulfuric acid

sulfur dioxide + air

absorption of sulfur dioxide

contact oven

roasting oven

iron oxide

The soaps that we use on a daily basis are products of the chemical industry.

EXPLOSIVES

The oldest of the explosives is **black powder**, which is made by mixing **potassium nitrate**, **charcoal**, and **sulfur**. The charcoal and sulfur are ground together in a large drum containing heavy steel balls, while the potassium nitrate is milled separately. The three components are mixed together and pressurized to form a paste. The finished powder exhibits glazed granules, which are produced by passing it through **wooden cylinders** in constant rotation. The **pyrotechnics** industry uses the powder and other **mild explosives**, to which mineral salts and other products are added, to create wonderful **fireworks**.

SOAPS AND DETERGENTS

One of the most important branches of the **light inorganic** chemical industry is the one that produces soaps and detergents. In order to be effective, these products have to be able to **capture** particles of **dirt**, form a **film** to keep them from reattaching to the **cleaned** surface, and **convey** the dirt to the **surface** of the water. The main raw materials that are used for making soaps are **fats**, such as **olive oil**, and **alkalis**, such as **sodium hydroxide**. They are ground up together, **glycerine** and **water** are added, and the mixture is subjected to high pressure and temperature.

INORGANIC FERTILIZERS

Intensive use of agricultural soil depletes it, so it is necessary to add significant quantities of **nitrogen**, **phosphorus**, and **potassium**. The most important fertilizers are the **nitrogenated** ones, which are made from **ammonia** and **nitric acid**. Ammonia is formed by the reaction of **nitrogen gas** and **hydrogen gas** in a **synthesizing oven** at high **pressure**. Nitrogen is normally produced by the combustion of **coal**. Hydrogen is obtained from the decomposition of **water** or from **petroleum**.

Fireworks are manufactured by the pyrotechnics industry.

Although it is believed that the Chinese invented gunpowder, it was the Arabs who first used it in war in the thirteenth century.

THE HEAVY PETROCHEMICAL INDUSTRY

The **organic** chemical industry has experienced a tremendous growth in the past 40 years, and its finished products include everything from countless **plastics**, **solvents**, **paints**, **synthetic rubbers**, and **explosives** to **textile fibers**, **fertilizers**, **alcohols**, **cosmetic perfumes**, and **pharmaceuticals**. **Xylene**, **toluene**, and especially **benzene** are products of the **fractional distillation** of **petroleum**; they are used as **raw materials** in the manufacture of many of the products mentioned previously.

LAMP BLACK

This is an important ingredient in **tires**, **printing ink**, and **paints**. At first this was obtained as a by-product of the incomplete **combustion** of **hydrocarbons**, but today it is produced by heating a hydrocarbon in **ovens** to separate the **hydrogen**, which oxidizes, from the **carbon**, which is extracted from the oven in the form of a fine powder.

HYDROCARBONS

In addition to petroleum, there are other hydrocarbons such as **natural gas** that are sources of raw materials including **methane**, which is used in manufacturing such varied and important products as **carbon tetrachloride**, a solvent, **nitrogen fertilizers**, **lamp black**, **ammonia**, and **anesthetics**.

Nitrogen fertilizers manufactured by the chemical industry are used to enrich large agricultural fields.

ALCOHOLS AND THEIR DERIVATIVES

Industrial alcohols are used in the production of **formaldehyde**, a reagent with many uses, various plastics, and **methyl chloride**, which is a refrigerant. **Methanol** is the most important industrial alcohol; it is obtained by causing a reaction between **carbon monoxide** or **dioxide** with **hydrogen**, all of which are in a gaseous state. **Ethanol** is widely known as a component in distilled and fermented drinks such as beer and wine, and it is also an important raw product in producing **plastics** and **acrylic fibers**.

Many articles of clothing that we wear are made of synthetic fibers. The photo shows a textile factory in Germany.

The chemical industry also produces perfumes, which are made chiefly from extracts of certain flowers, leaves, and fruits.

Although most perfumes are animal or vegetable in origin, the chemical industry can also produce them synthetically from coal tars.

The alcohol contained in beer is known as ethanol; unlike other alcohols, it can be ingested.

PLASTICS

This term covers a wide variety of materials that are generally produced by **polymerization**, in other words, by joining several simple **molecules** to form a single **huge** one. Some plastics are molded by injecting them into **molds**, where they harden, and others by **punching** them out in a rigid state using a press. Some are **fibers** that are meant to be woven, and others are **condensed** into flexible, transparent sheets; but all of them are dependent on **parkesine** or **xylonite**, of which the most widely known derivative is celluloid, which was invented by Alexander Parkes in 1862.

NATURAL POLYMERS

Although the word *plastic* is sometimes used as a synonym for artificial or unnatural, some plastics are produced from materials as natural as **wood** and **marine algae**, since they both have large molecules that are easy to polymerize. Products that are produced in this way include **rayon**, **acetate**, **caseine**, and **alginate fiber**.

MAKING PAPER

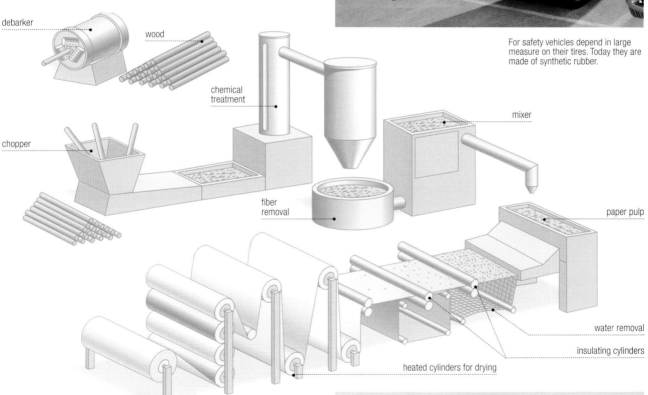

debarker
wood
chemical treatment
chopper
mixer
fiber removal
paper pulp
water removal
insulating cylinders
heated cylinders for drying

There are many different qualities of paper based on their intended use. Paper used for books is normally high-quality art paper.

SYNTHETIC RUBBER

In 1826 Michael Faraday discovered that natural rubber is a hydrocarbon with a short chemical chain involving five atoms of **carbon** and eight of **hydrogen**, and that was the beginning of efforts of produce it artificially. Today there are many types of synthetic rubber that are difficult to classify, because often the name given them is a commercial brand. The best-known one is **styrene-butadiene**, which is produced by mixing the two components in a soapy, watery solution that is beaten to form an emulsion, which is synthetic rubber.

For safety vehicles depend in large measure on their tires. Today they are made of synthetic rubber.

PAPER

The raw material used in making paper is **wood pulp** or **used paper**. The most current method is the **Kraft** process, which involves putting the clean, chopped pulp into a **digester** under elevated **temperature and pressure** where it is treated with **caustic soda** and **sodium sulfate**. The process is completed by **washing**, **screening**, **drying**, **filtering**, and **bleaching** the various intermediate products until the desired sheets of paper are obtained.

CONSTRUCTION MATERIALS

For thousands of years, the construction materials used for every area depended on the climate, the available materials, and people's lifestyle. At first, this involved natural products used as they were found, but little by little they underwent simple modifications.

When commercial and cultural exchanges became feasible, some populations introduced innovations that were adapted to the lifestyle, but it was not until the nineteenth century that industrialized, urbane western culture imposed a nearly universal construction model.

PRIMITIVE MATERIALS

In the early millennia, humans, who were entirely **nomadic**, had to depend on shelter offered by the terrain when it came to spending the night; they could make conditions more comfortable by gathering **leaves and branches**, much as chimpanzees and gorillas do. The people who learned to follow flocks of animals discovered how to use **hides** to construct light, comfortable shelters. Some cultures that practice **itinerant livestock raising** and have survived into modern times combine **skins**, **flexible branches**, and **cloth** to make attractive and varied **temporary** shelters that are easy to dismantle.

HUTS AND IGLOOS

Some pastoral communities constructed semipermanent shelters by **packing** a framework of **branches** with a mixture of **earth** and **animal dung** blended to a paste with water, which hardened as it dried. **Sticks** and **skins** were used for the roof and the door, as well as for lining the interior. Nothing but blocks of **ice** are used by Eskimos for constructing their **igloos**, which protect them against the cold; the interior may be 32°F (0°C), while outside it is minus 25°F (minus 30°C) or colder.

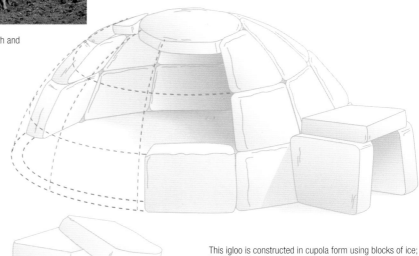

As early as thousands of years ago, people were constructing light tents of hides and poles. They could be quickly set up and taken down and moved to areas where the hunting was better.

The huts of the Masai in Africa are made from a mixture of earth and manure; they provide good protection from the torrid sun.

ROOFING

The materials used to cover dwellings may be of **plant origin**, such as straw, palm leaves, dried branches, wood, or thatch; of **mineral origin**, such as slate; of **animal origin**, such as hides; or **man-made**, such as bricks, cloth, and metal sheets.

This igloo is constructed in cupola form using blocks of ice; the blocks fuse together upon contact.

THE FIRST DURABLE MATERIALS

Stone and **compressed earth** were surely the first durable materials used in construction. In making the most primitive stone houses a hole was first made in the ground and **surrounded** with piled-up **rocks**. These dwellings rarely exceeded 3 feet (1 m) in height; there were no windows, and the roof consisted of branches held in place by rocks. Houses made of packed dirt were constructed by pouring between two horizontal logs or planks a mixture of **clay-rich earth** and **straw** brought to a doughy consistency by adding water and stamping on it to compact it and remove excess water. This system makes it possible to construct large houses with windows, but it requires very thick walls when beams are used for the roof.

There are two types of construction in Yemen (Asia Minor): adobe (at right), used in hot regions, and stone (left), used in mountainous and rainy areas.

ADOBE AND BRICK

Adobe is a variant of compressed earth that involves making medium-sized blocks using the same materials and letting them **harden** in the sun. The blocks are held in place with the same paste before it dries, to which lime is sometimes added. The practice of **baking** blocks of **clay** to harden it is very ancient; it was used more than 6,000 years ago in Asia for making bricks. The combined use of **bricks** and **wooden beams** made it possible to construct houses of more than a single story, with several rooms, doors, and windows.

COMPLEMENTARY MATERIALS

When people began to construct large houses that required **windows**, it became necessary to find a material that would allow the passage of **light** but keep out **water** and **wind**. **Parchment** was used in some places; this is a skin that has been hardened and worked until it becomes **translucent**. In other places **papyrus** was used; this is a type of tough paper made from the fibers of the plant that has the same name. Planks of **smoothed wood**, flat flagstones, and baked clay tiles were used for the floor; however, it was more common to cover the floor with carpets or mats woven from plant fibers.

Because it is easily worked, wood has been used for construction throughout the ages. Its greatest drawback is its combustibility.

Brick has come into increasing use in the last 5,000 years. It is generally used for construction (at left, the house of Saint-Bernard-de-Comminges in France); however it is also used as an architectural and artistic material (at right, the Romanesque tower of Saint-Sernin in Toulouse, France).

NEW MATERIALS

One of the limiting factors in stone and brick construction was that the mixes used to hold the blocks together were not strong enough, and over time they lost their consistency and deteriorated. The best results were obtained by mixing carefully selected **sand** with **ground limestone**. The **ancient Romans** added **gravel** to this mix to produce **concrete**, which they poured into **molds** while it was still liquid, in order to make **beams**, **columns**, and other elements that allowed construction of higher, safer, and better illuminated buildings.

PORTLAND CEMENT

If the pulverized mixture of sand and limestone is **calcined** in ovens at 2,400 to 2,550°F (1,300 to 1,400°C) and small quantities of **gypsum** are added, the result is **portland cement**, which is used universally in modern construction. Today the sand used in making this cement is carefully selected so that it contains the precise quantities of **iron**, **aluminum**, and **magnesium** that every type of cement needs.

Large works of modern engineering require strong and durable materials. The photo shows the Cahora Bassa dam in Mozambique.

CONCRETE

Toward the end of the nineteenth century it was discovered that **portland cement** becomes much more resistant to **traction** if **iron mesh or rods** are added to it. This new material, which is called **reinforced concrete**, revolutionized construction methods in the following century; it made it possible to construct **skyscrapers** and **dams**, plus impressive **bridges** and **viaducts**, among other large projects. There are several types of concrete that are distinguished by their components, depending on the use for which they are intended.

GLASS

During the Middle Ages only important buildings such as palaces and cathedrals had glass windows, for at that time glass was made only in small quantities and was too expensive for use in everyday buildings. In modern glass production, a mixture of **sand** and **sodium carbonate** or **sulfate** and **lime** is **melted** and cooled quickly to avoid crystallization; however, the tremendous variety of glass that is produced today means that this formula is merely representative.

PROCEDURE FOR MANUFACTURING FLOATED GLASS

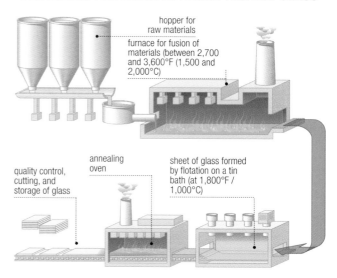

hopper for raw materials

furnace for fusion of materials (between 2,700 and 3,600°F (1,500 and 2,000°C)

annealing oven

sheet of glass formed by flotation on a tin bath (at 1,800°F / 1,000°C)

quality control, cutting, and storage of glass

Floated glass refers to flat panes of glass used in large buildings and the windows of private houses.

The first type of glass used in construction was the stained glass in the windows of palaces and cathedrals.

METALS USED IN CONSTRUCTION

Although metals were used even in very ancient times as **decorative** elements in important buildings, they have been used as common construction materials only for a relatively short time. In addition to using **iron** in concrete, the widespread incorporation of sanitary facilities inside homes made **lead** the ideal material for conveying water; **copper** was soon added for use in **electrical conduits** and pipes for **hot water**.·

The construction of large buildings requires the use of steel frames.

Steel, glass, and aluminum are commonly used for the finishing touches of large skyscrapers.

STEEL

When it became possible to produce steel industrially at a reasonable price, it came into broad use in construction. Many modern **skyscrapers** have a steel **framework**, but the most visible use of steel is in large **bridges**, especially in **suspension** and **drawbridges**. Steel brought **strength** and **lightness** to construction, but it requires periodic maintenance to prevent damage from **corrosion** caused by the environment.

THE GROWTH OF ALUMINUM

Reduced production costs and the discovery of new alloys have made aluminum a serious competitor for steel in the construction industry. Certain aluminum **alloys**, which have been used for some time in making **large windows** and as a covering for **facades**, are now being used as substitutes for steel in large building projects, since they are **lighter** and just as **strong**.

OTHER MODERN MATERIALS

Plastics occupy an important position among modern construction materials, especially for producing pipes for plumbing, where PVC is a material that is used universally. It is also an important material in **thermal and acoustic insulation**; along with **fiberglass** it is the most commonly used material.

Fiberglass is a light and very strong material that can be used to make any number of items. The photo shows a windsurfing board made of fiberglass.

PUBLIC WORKS

A constant desire of humanity has been to provide food, comfort, and security with minimal effort, using technology as the main implement; however, none of that would be possible if people were not able to organize into stable societies capable of working together. Population growth makes certain works and services indispensable, as well as the technical knowledge and economic, social, and political systems that make them possible.

ROADS AND HIGHWAYS

The first highways came into being at the time of the great **empires** of antiquity, and they don't postdate the domestication of the **horse** and the invention of the **wheel** by very much. The roads constructed by the Romans more than 2,000 years ago were notable for both their ingenious **layout** and the **quality** of their construction. The Appian Way, the construction of which began in 312 B.C. and ended up being more than 400 miles (660 km) long, had a consistent width of 37 feet (12 m), of which 32 feet (10.5 m) were designed for horse and carriage traffic and paved with flat stones.

MODERN HIGHWAYS

Roads designed exclusively for automobiles were not constructed in Europe and America before 1910, and the first highways date from 1930. When highways are built, engineers consider not only the places that need to be connected, but also the **topography** of the land in order to avoid as much as possible major **changes in elevation**. The majority of the money that is spent on highway construction involves **earth moving** to level out the route.

STRUCTURE OF THE PAVEMENT ON ROMAN ROADS

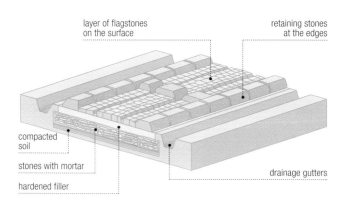

layer of flagstones on the surface

retaining stones at the edges

compacted soil

stones with mortar

hardened filler

drainage gutters

PAVING

Once the terrain has been leveled, the surface on which the vehicles will travel is constructed. Layers of **sand and gravel** or crushed rock are put down; in some cases these are covered with portland cement and then with **asphalt**, a black, sticky petroleum derivative that is applied hot in successive layers that may total 8 inches (20 cm). Modern asphalt is prepared in different mixes based on the **climate** of the area and the **intensity of the traffic** it has to bear; there are even some types that can **absorb** some of the **noise** produced by vehicles.

Current appearance of the Roman road that joined Ostia with Rome; it was constructed more than 2,000 years ago.

The tremendous number of vehicles that circulate around large cities have made it necessary to construct major highways and interchanges such as this network of highways near Chicago.

BRIDGES AND VIADUCTS

The solution to constructing a short bridge on a local road used to involve tree trunks or flat stones, but for busy roads and long bridges, the structures have to be **secure** and **long lasting**. Three types of solution have been used since ancient times: **arch** bridges, bridges supported by **girders**, and **suspension** bridges, each of which has advantages and disadvantages depending on the nature of the location.

There are other types of bridges, such as mobile ones, which are constructed over navigable waterways; they allow ships to pass when part of the bridge retracts (either by raising or sliding).

TYPES OF FIXED BRIDGES

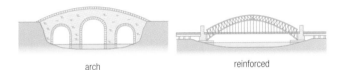

arch   reinforced

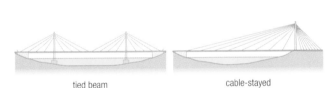

cantilever   suspension

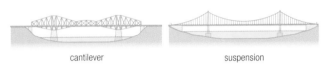

tied beam   cable-stayed

The old Tower Bridge in London is a drawbridge; the road splits in two to allow ships to pass.

BRIDGES AND VIADUCTS ON GIRDERS

The main characteristic of bridges on girders is that all of their weight rests on the **columns** that support them. Although their construction is often **inexpensive**, they must stand on very **firm** ground, and water flow should remain **steady**. When these factors are coupled with uneven terrain over which water cannot flow freely, conditions are perfect for constructing **viaducts on girders**. At present, parts are assembled at the construction site using **prefabricated** components.

 Some bridges are real technical challenges. The traveled portion of the Honshu-Shikoku Bridge (Japan, 1998) is over 1 mile (2 km) long and is held up by steel cables that are 43 inches (110 cm) in diameter.

ARCH BRIDGES

Arches spread out the weight **vertically**, onto the support **pillars**, and **horizontally** onto the contiguous arches and the **supports** of solid ground; this allows for longer spans between pillars than with bridges made with girders. Some examples of **Roman** and **medieval** arch bridges still exist; they were constructed from stone and brick. **Iron and steel**, beginning in the twentieth century, made it possible to make bridges of this type that are very long and elegant. Many modern bridges combine arches with the technique of suspension bridges.

SUSPENSION BRIDGES

With suspension bridges, the weight falls on the anchor points of the cables that hold them up, or on the anchor points and the **vertical supports** when they contact solid ground. The **traveled surface** of the bridge is held up by **cables** that are secured to the upper part of the **towers** and whose ends are anchored in the **support pillars**. The main advantages of this type of bridge are that the traveled portion of the bridge can be suspended at **considerable height**, and the **pillars** can be very far apart.

In addition to their primary function, cable-stayed bridges are esthetically very attractive.

TUNNELS

The practice of constructing tunnels to simplify laying out highways and railroads began at the same time that **railroads** were built. The first railroad tunnels were dug by hand using knowledge from mining, and they cost many lives. Today, tunnels are dug using **specialized machinery** and very safe methods.

WAYS OF DIGGING

Tunnels are dug simultaneously from both ends by using **explosives** and **excavators**. The traditional methods, which are known by the names of the countries where they were first used, are the **English**, which involved digging **horizontal** strips the entire width of the tunnel from top to bottom; the **German**, which involved digging the **sides** and leaving the center **intact** until the end; and the **Austrian**, which began with a gallery at the **base** of the tunnel and a second one dug **higher up** and **joined** with the first one by means of **chimneys**. Modern excavators dig the entire width and height of the tunnel, and the **prefabricated support elements** are put into place behind it.

Tunneling machines bore through the subsoil with greater speed and safety than traditional diggers.

PORTS

The first ports contained a single **dock** for mooring made of stone or wood, located on protected parts of the coast such as deep bays. As maritime transportation became more important, docks were added, as were such installations as **lighthouses**, **warehouses**, and other **ancillary buildings**. Coastal cities that were not favored by geographical chance had to construct artificial bays using **breakwaters** made of enormous stone blocks covered with concrete to create an inner area of **calm waters** where **docks** could be built.

MODERN PORTS

Modern ports are characterized by **specialization**; there are specific docks set aside for traffic relating to **oil and petroleum derivatives**, moving **containers**, and passengers. The ports also use spectacular **machinery** for loading and unloading ships, and they have a broad variety of installations and services.

Two sections of the Barcelona (Spain) port: above, the tourist terminal; left, the container terminal.

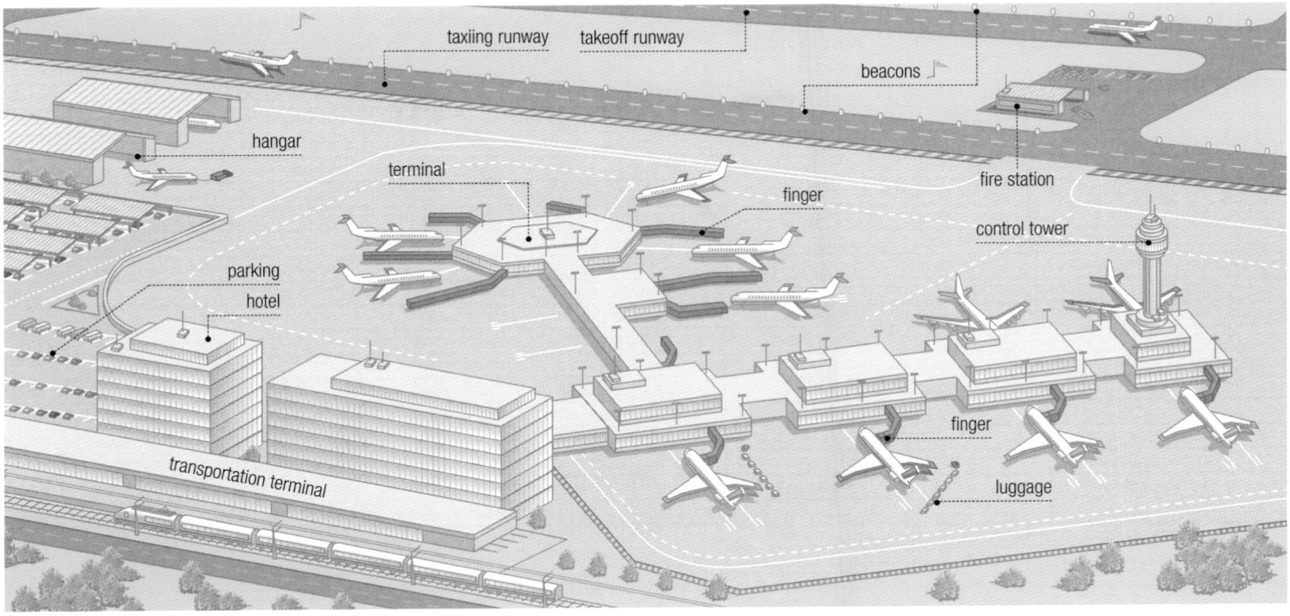

The image diagram labels:

taxiing runway takeoff runway

beacons

hangar

terminal

finger

fire station

control tower

parking

hotel

finger

luggage

transportation terminal

AIRPORTS

The first airplanes did not need special provisions for landing and taking off; all they needed was a flat area a few dozen yards (meters) long. However, the tremendous development that aviation experienced starting in 1920 made it necessary to construct proper areas for these functions. The first airports were adequate only for day flights, for aside from the landing strip the installations were very rudimentary.

The complexity of air traffic requires a highly dependable control system. The photo shows the interior of the control tower at the airport in Washington, D.C.

RAILROADS

In laying out rail lines the grades are kept to 5 percent or under when they are to be used by high-speed trains, so **grading** becomes very important, as does the construction of **bridges**, **viaducts**, and **tunnels** to simplify the layout. **Curves** have to have a very gradual **radius** so that trains can negotiate them without jumping the tracks.

CONSTRUCTING THE RAIL LINES

Ballast is spread onto the ground; this consists of hard rocks 1 inch or 2 (3 to 6 cm) in diameter that provide a base that is **elastic** but **firm** to **spread out** the **weight** of the trains over a broad surface and allow for the drainage of rainwater. The steel **rails**, which are like an I-beam in cross section, are secured by railroad spikes to the regularly spaced **ties**.

MODERN AIRPORTS

Modern international airports have several takeoff and landing runways, some of which are almost 2 miles (3 km) in length; they are perfectly marked and made to function in all kinds of weather. The main installation in an airport is the **control tower**, where the **air traffic controllers** watch over all the movements of the planes in the vicinity by means of **radar** installations, and they stay in contact by **radio** with the planes that are ready to take off or land. The **terminals** are the increasingly large buildings where the various services are concentrated that need to be provided to the throngs of travelers who use them.

The restricted maneuverability of trains requires the construction of huge viaducts and bridges.

Rails of a cog railway resting on ballast (the stone base).

WATERWORKS

Water is an indispensable element for life, and humans have realized this since the earliest times. **Canalization** of certain rivers and streams was among the first projects carried out by humans for watering crops; the periodic floods of others served the same purpose. As populations grew, it became necessary to devise ways to get water to the new population centers and their agricultural fields.

Irrigation canals make it possible to get water to good agricultural areas where there is insufficient rain.

IRRIGATION SYSTEMS

It is not always possible to use the water from rivers by simply **digging** a **network** of canals to distribute the water; sometimes it is necessary to **raise** the watercourse to remedy problems with the terrain. **Waterwheels**, moved by humans or animals, were one of the earliest mechanisms used to solve this problem. The great **seasonal** fluctuations that some rivers experience, in which the flow rate is **lowest** during the **greatest need** for irrigation, was partially solved by constructing dams.

For many centuries waterwheels were used to raise the water from a river or a canal to a higher level and make it easier to transport.

DAMS

Dams are constructed to hold back part of the water in rivers at the time when they experience their greatest flooding so that the water can be used when it is needed. The oldest and simplest use earthworks, and the flow of water is hindered by an **accumulation** of **dirt** and other materials on a very **broad base** that **narrows** toward the top. The earliest known dam of this type was constructed in the Nile valley some 5,000 years ago.

TYPES OF DAMS

In addition to earthwork dams, other types are constructed, such as **gravity** dams. The term comes from the fact that it is the **weight** of these dams that keeps the earth on which the **retaining wall** rests from slipping; the wall is vertical **upstream** and **broadens** to form a **triangular** cross section on the downstream side. There are also **arched** dams that present a **curved** wall to the water, thereby **diverting** part of the pressure toward the sides; as a result, they do not need to be as thick.

MAIN TYPES OF DAMS

gravity dam

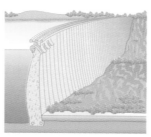

arch dam

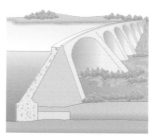

buttress dam

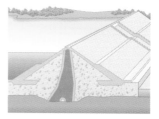

earthwork dam

NEW USES FOR DAMS

Modern dams do more than store up water; they are also an important source of **energy**, since they produce a major part of the **electricity** that we use. The **force** of the escaping water is used to drive the **hydraulic turbines**, which in turn move the electrical generators. When the dammed-up water covers a **large area** it is also used to construct **sporting** and **recreational** facilities.

 The highest dam in the world is the Rogunsky dam in Tajekestan, at 1,038 feet (335 m); the longest is the Yacyreta-Apipe dam on the border between Paraguay and Argentina, which is almost 42 miles (70 km) long.

Despite their incomparable benefits, large dams have a profound impact on the countryside.

NAVIGABLE CANALS AND LOCKS

Until the railroad came into general use, most **goods** were transported on **waterways** in countries that were blessed by geography, taking advantage of rivers and an extended network of **navigable canals** that had been constructed through the centuries. These networks, which sometimes connect the navigable stretches of various rivers, are possible only through the utilization of **locks**, which make it possible to reconcile the differences in altitude among rivers.

DIAGRAM OF HOW A LOCK WORKS

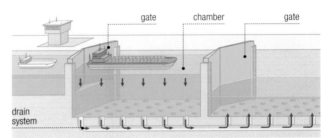

1. Once the ship is inside the chamber, the water is let out through the drain system.

OCEAN CANALS

Ocean canals are artificial watercourses that join two oceans or seas to avoid long **detours** or **dangerous crossings** during certain times of the year. Some ocean canals use **several locks** to handle the differences in altitude in a number of steps. The longest one is the **Suez Canal** in Egypt, which is 103 miles (169 km) long; it links the Red Sea with the Mediterranean. The next longest are the **Kiel Canal** at 58 miles (95 km), which makes it possible to get from the Baltic Sea to the North Sea, and the **Panama Canal**, which is 52 miles (85 km) long.

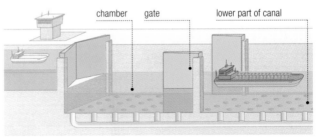

2. When the water of the chamber reaches that of the canal, the gate is opened and the ship moves on.

HOW A LOCK WORKS

A lock is a stretch of **canal** provided with watertight gates that can be **filled** with water and **emptied** at will. When a ship enters the lock the gates are closed and the water level is brought up to the level of the canal through which the trip will continue. Once this operation is complete, the **forward gate** is opened and the ship continues on its way.

The Panama Canal links the Atlantic Ocean with the Pacific. It was very difficult to construct because of the geographic obstacles.

TRANSPORTATION VEHICLES

People's transportation needs are closely linked to the level of development of the communities where they live. When people need to exchange products that they don't consume themselves, they have to develop increasingly efficient and safe means of transportation, but this desire has encountered natural obstacles that science and technology have spent a long time overcoming.

LAND TRANSPORTATION

Sleds were useful for transporting awkward and heavy objects, but they could be used only on fairly **flat** areas and on surfaces that facilitated **sliding**, such as earth or packed snow. Their serviceability increased somewhat when **domestic animals** were used to pull them, but until the **wheel** was invented their use remained very limited.

Sleds glide on a pair of runners and are usually drawn by reindeer or dogs. Today, they are used principally for recreation.

WAGONS

Before the wheel came into use, it is probable that two tree trunks were suspended by leather thongs under sleds to make them easier to move, but as early as 3500 B.C. the **Sumerians** had actual carts with solid wheels that were an improvement over sleds. Variations of the Sumerian cart appeared in the following centuries until around 2000 B.C., when the **Hittites** constructed **light wagons** with **spoked** rather than solid wheels, which could be pulled by recently domesticated **horses**.

The great ancient civilizations, except for the pre-Columbian ones, made use of the cart.

CARTS AND COACHES

Two basic types of wagons—**carts**, which were slow but strong, for transporting **heavy** objects and generally pulled by oxen, and light, horse-drawn **coaches** for transporting people—were the standard land transportation vehicles for centuries, and the improvements that were made applied mainly to the **harnesses** for the horses and the **suspension** for the riding compartment or the cargo bed on the **axles**.

The oxcart is more than 4,000 years old and is still used in many parts of the world.

STAGECOACHES

The stagecoach is one derivative of the wagon, but its main novelty consisted of using a suspension system that made travel over rough roads more comfortable.

Stagecoaches did not contribute great technological innovations, but they ushered in the concept of **public transportation**. Stagecoaches, which appeared at the start of the seventeenth century, were heavy carriages pulled by a team of from two to eight horses. They kept to **fixed** routes and schedules and were a means of transportation for people who did not own a horse or who were afraid to take certain dangerous trips alone.

BICYCLES

In 1839 the English blacksmith Kirkpatrick Macmillan constructed a two-wheeled vehicle whose rear wheel was larger than the front, with a saddle between them, and that could be propelled by pedaling. It was the first bicycle, heavy and ungainly, from which modern bicycles evolved.

THE COMPONENTS OF A BICYCLE

mirror

saddle

tail light

rear brake

reflector

brake lever

handlebar

frame

headlight

front brake

generator

fork

derailleur

chaining

pedal

chain

spoke

tire

rim

THE RAILROAD

Starting with the invention of the **steam engine** many inventors saw the possibility of converting it to the motive force of **wheeled vehicles**, and as early as 1771 there were prototypes that made use of this idea. The Frenchman Nicolas Joseph Cugnot constructed a steam-powered **tricycle** that could move at about 4 miles (6 km) an hour, and a few years later the British inventor Trevithick got the idea that he could improve the vehicle's performance and make it move more easily by having it travel on iron rails.

THE FIRST LOCOMOTIVES

In 1804 Trevithick made a locomotive that was capable of pulling five loaded wagons for 9 miles (15 km) at a speed of almost 5 miles (8 km) an hour, but he didn't succeed in securing funding to exploit his invention commercially. In 1825 another Englishman, George Stephenson, was more successful when one of his locomotives pulled 38 cars at the speed of over 12 miles (20 km) an hour between the English towns of Stockton and Darlington.

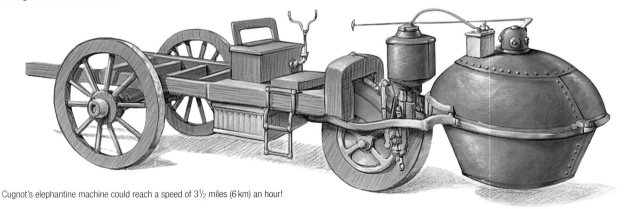

Cugnot's elephantine machine could reach a speed of 3½ miles (6 km) an hour!

THE EVOLUTION OF THE RAILROAD

It is not surprising that Stephenson's invention became a success, for that was the first time that the **speed of a galloping horse** had been exceeded over a long distance; it also provided greater **security** for passengers and goods than the types of land transportation that had been used previously—**stagecoaches** and **caravans** of draft animals. In its first 100 years of existence, more than 600,000 miles (a million km) of rail were laid, and this helped unify large countries such as the United States and Russia.

Today, old steam engines are still in use for the pleasure of railroad fans.

HIGH-SPEED TRAINS

Competition from the airline industry and improvements in roads resulted in a certain stagnation in the railroads between 1950 and 1980, but at least in Europe, the latter date marked a new start thanks to the appearance of **high-speed trains** fitted with the greatest comforts and capable of exceeding **180 miles** (300 km) an hour in **complete safety**. In addition to using high technology in the rolling stock, these types of train need **special rails** with very **gradual curves** and **minimal grades**.

NEW LOCOMOTIVES

Starting with Stephenson's locomotive, in a few decades, during World War I, there were others that could move 600 tons at 72 miles (120 km) an hour; however, steam locomotives soon began to be replaced by **diesel** locomotives and **electric trains** that provided more power with less pollution.

SPECIAL TRAINS

Conventional railroads have certain limitations in the grades that they can climb, but these difficulties were resolved by inventing new **traction** systems. **Cog railways** use a third **geared wheel** that meshes with the rack located in the **center of the track**; **funiculars** go up **short** tracks on **steep grades** by means of a cable that pulls them up. Funiculars almost always work in pairs that are connected by a cable in such a way that the weight of the car going down helps raise the other one.

Cog railways use a geared central wheel for ascending steep grades.

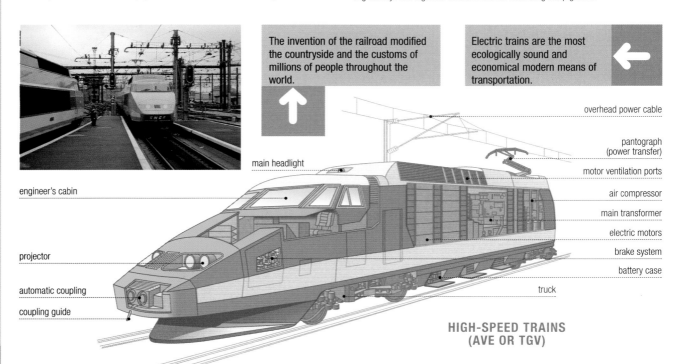

The invention of the railroad modified the countryside and the customs of millions of people throughout the world.

Electric trains are the most ecologically sound and economical modern means of transportation.

engineer's cabin

main headlight

projector

automatic coupling

coupling guide

overhead power cable

pantograph (power transfer)

motor ventilation ports

air compressor

main transformer

electric motors

brake system

battery case

truck

HIGH-SPEED TRAINS (AVE OR TGV)

WATER VEHICLES

Rafts made of logs tied together with **ropes** or vines are probably at least as old as the most primitive sleds, and the same can be said of the first **canoes**. Among these the **kayak** or Eskimo canoe deserves special mention; for lack of wood, these craft were constructed using whalebones and sealskins stitched together. There is evidence that around 3500 B.C. solid, graceful, flat-bottomed ships powered by oars or sails were traveling on the navigable section of such rivers as the Nile, the Tigris, and the Euphrates; these craft were capable of transporting people, animals, and merchandise.

OCEANGOING NAVIGATION

In very early times some of the best river vessels probably ventured out of the protection of the large river deltas and traveled along the coast. But true oceagoing navigation was started by the **Phoenicians**, who took their bearings from the seven stars that make up the constellation **Ursus major**, which is always located in the north; around 1100 B.C. they ventured forth to explore the Mediterranean.

The ancient kayaking practiced by Eskimos has been transformed into an Olympic event.

PHOENICIAN AND GREEK SHIPS

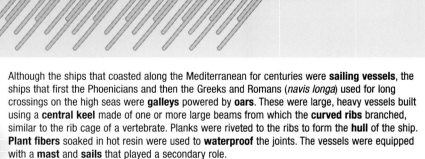

Galleys were propelled by one, two, or three rows of oars on each side of the ship. It was very difficult to operate them, and the hard labor was generally done by slaves or prisoners.

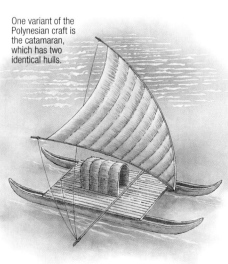

One variant of the Polynesian craft is the catamaran, which has two identical hulls.

Although the ships that coasted along the Mediterranean for centuries were **sailing vessels**, the ships that first the Phoenicians and then the Greeks and Romans (*navis longa*) used for long crossings on the high seas were **galleys** powered by **oars**. These were large, heavy vessels built using a **central keel** made of one or more large beams from which the **curved ribs** branched, similar to the rib cage of a vertebrate. Planks were riveted to the ribs to form the **hull** of the ship. **Plant fibers** soaked in hot resin were used to **waterproof** the joints. The vessels were equipped with a **mast** and **sails** that played a secondary role.

THE POLYNESIAN CATAMARAN

This is a sailing vessel that has a second hull to provide stability. Using catamarans, the ancient Polynesians managed to conquer hundreds of the islands scattered over thousands of miles (kilometers) in the Pacific Ocean.

A REIGN OF 2,000 YEARS

The Phoenician and Greek galleys were the model for the ones that the Romans and the Carthaginians used to consolidate their huge empires. With hardly any improvements they were the **military** and **commercial** ships that dominated the Mediterranean, the Baltic, the North Sea, and the northwest Atlantic until the thirteenth century of the Christian era, when a ship that was developed primarily by **Portuguese** sailors came not only to dominate oceangoing navigation, but also the face of the known world.

THE CARAVEL

The **caravel** was a three-masted vessel that was driven solely by sail, could resist **heavy seas**, and was very maneuverable. Portuguese and Spanish sailors used the caravel for increasingly daring voyages on the Atlantic. The Portuguese accomplished the first important feat in 1418 when they discovered and colonized the island of **Madeira**, followed by the discovery in 1427 of the archipelago of the **Azores**, located 720 miles (1,200 km) from the Portuguese coast. The only navigational instruments they had at their disposal were the **astrolabe**, which was useful for verifying the **latitude** of the sun and the stars, and the **compass**, which had been invented in China and adapted to **navigation** by the **Arabs** around the twelfth century.

For many centuries it was easier and less expensive to do business with ports located thousands of miles (kilometers) away than with inland cities separated by a couple of hundred.

The caravel played an important role between the sixteenth and the seventeenth centuries, when it was replaced by galleons, which were larger and more secure.

TRANSATLANTIC NAVIGATION

In 1492, when Christopher Columbus reached the coast of the Americas, he not only put two continents in touch that were separated by about 4,800 miles (8,000 km), but he also launched a new era in which oceangoing transportation would be the main feature. The **frigate** would become the most representative of the large **sailing ships** that would furrow all the seas and oceans in the following centuries. The frigate had three masts that were taller and rigged more thoroughly than the caravel, and as a result it spread a **greater area of sail** to the wind and was capable of carrying up to **700 tons** of goods in its **hold**.

CROSS SECTION OF A WARSHIP FROM THE MID-EIGHTEENTH CENTURY

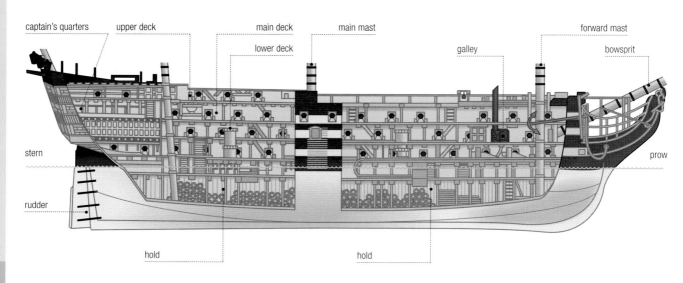

captain's quarters — upper deck — main deck — main mast — forward mast

lower deck — galley — bowsprit

stern

rudder

prow

hold   hold

STEAMSHIPS

At the start of the nineteenth century the **steam engine** was adapted to powering ships, especially riverboats. The energy contained in the steam was used to move one or two **wheels** fitted with **paddles** that provided a constant **navigational speed**, independently of the speed and the direction of the wind. Oceangoing navigation did not benefit very much from this system, since the paddle wheels weren't very effective in **strong seas**; however, in 1827 the British engineer Robert Wilson developed a **propeller** that pushed the ship from the stern and was not affected by the waves.

View of the drive wheel of the *Natchez*, a steamboat that plied the Mississippi in the middle of the nineteenth century.

IRON HULLS

The installation of steam engines to provide the motive power in ships and improvements in the iron industry made possible the gradual **replacement** of wooden hulls by **iron hulls**; properly protected against rust by painting, these offered greater security and made it possible to construct increasingly large vessels. The **structure** used in **iron hulls** is similar to that of **wooden** ones, but the way it is made is totally **industrial**, as compared to the hand work used in wooden ships.

MODERN SHIPYARDS

The plant where ships are constructed is known as a **shipyard**. They are commonly located on the coast or in the mouth of a large river; that is where some of the ship components are manufactured, and the ones that are made elsewhere are assembled. The hull is constructed inside a **dry dock** called a **slipway**, which is located below water level. This is filled with water when the hull is finished so it can be towed to the **fitting dock**, where the engines and other components are installed.

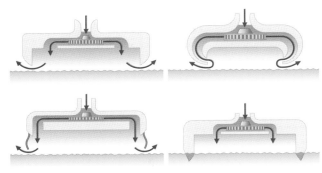

The principle of how the hovercraft works involves injecting air underneath the hull so that it "floats" on a cushion of air. Propulsion is effected by means of propellers located underneath (the same ones that inflate the cushion) or on the outside (as if it were an airplane). The illustrations show four variants of hovercraft.

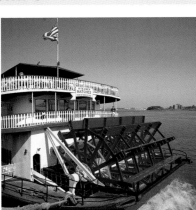

In the 1960s a revolutionary air glider known as the "hovercraft" made its appearance. This is an aquatic vehicle that can reach high speeds with remarkable stability.

SUBMARINES

Submarines are ships that are capable of sailing **under water** by means of a system of pumps that **fill** or **empty** large water tanks. They are a far cry from the prototype that a Spanish inventor named Narcís Monturiol tried out in the port of Barcelona in 1859. Modern submarines displace more than **10,000 tons**, are propelled by **nuclear reactors**, and can remain submerged for several months at a time.

PARTS OF A NUCLEAR SUBMARINE

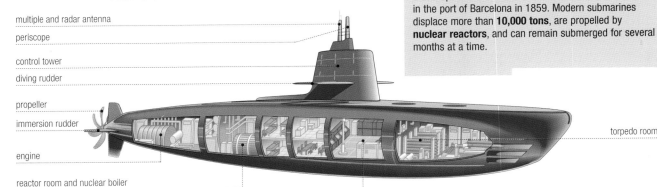

multiple and radar antenna
periscope
control tower
diving rudder
propeller
immersion rudder
engine
reactor room and nuclear boiler
control room
torpedo room

AUTOMOBILES

Even though we are accustomed to thinking of **automobiles** simply as the vehicles with which we are all familiar, the term really applies to all vehicles that are equipped with their **own means of propulsion**. In that sense, motorboats and airplanes are also automobiles, but the term has been restricted to vehicles that are used on solid ground. The main characteristic is that they have an **internal combustion motor** that usually burns fossil fuels.

Cadillacs coming off the assembly line in the Detroit factory in 1917.

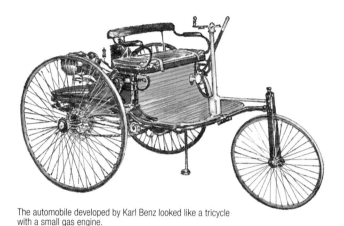

The automobile developed by Karl Benz looked like a tricycle with a small gas engine.

THE FIRST AUTOMOBILES

Even though the first automobile propelled by a steam engine was built in 1771, it wasn't until the end of the nineteenth century that the first truly useful autos were produced. The German engineers **Gottlieb Daimler** and **Karl Benz** independently built two vehicles that included the characteristics by which we recognize cars, and starting at that time many different models were manufactured in Europe and the United States. Some of them were real gems of hand-made mechanics, but these were very expensive vehicles available to only a few.

ASSEMBLY LINES

In 1910 the American engineer Henry Ford, who had several years' experience in manufacturing automobiles, decided to **mass-produce** them, in other words, to make identical copies of the same model on **assembly lines** where each worker performed the same operation quickly. This way of producing automobiles lowered their cost considerably and turned **Ford** into the premier world manufacturer during the 1920s. Ford's methods were applied throughout the world and in the following decades the auto was turned into an article of **mass production** accessible to almost everyone.

THE MAIN COMPONENTS OF AN AUTOMOBILE

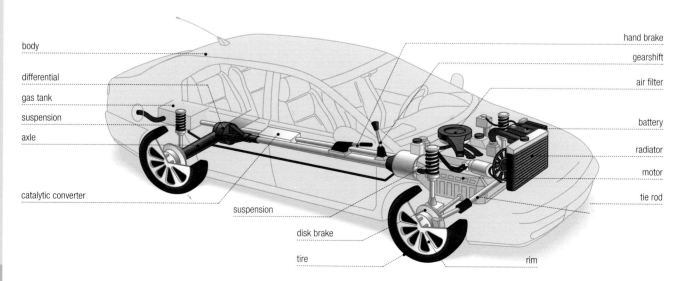

body
differential
gas tank
suspension
axle
catalytic converter
suspension
disk brake
tire

hand brake
gearshift
air filter
battery
radiator
motor
tie rod
rim

AUTOMOTIVE DESIGN

Based on the type of vehicle to be produced—family car, city car, sports car, forth—wheel drive, and so—the **space** is planned to allow for the greatest comfort for the driver and the passengers and leave a **cargo** area appropriate to the type of vehicle. The exterior shape corresponds not only to **esthetic** considerations, but it also presents the least possible **wind resistance**. **Weight distribution** is also important so that the car performs properly in curves without **rolling over**, as well as complying with all **safety regulations**.

INDUSTRIAL PRODUCTION

Presently, cars are produced in highly **sophisticated** plants, some of which are entirely **automated**, where some components are **manufactured** and the ones that are produced elsewhere are assembled. Before being used, all major components are subjected to **safety tests** established by regulations, and to others determined by the individual manufacturer.

headrest: prevents whiplash effect (violent jarring of the neck) in case of collision

seat belt: holds the body in place in case of collision

airbag: inflates if vehicle is involved in a collision

Crash-test dummy on a safety test bench.

THOROUGH TESTING

Throughout the various phases of their production, automobiles are subjected to different tests in addition to the ones applied to individual components. The performance of the body is tested in a **wind tunnel** that simulates driving conditions so that the wind currents around the vehicle can be studied. The **mechanical** performance is checked on closed circuits and the **safety systems** are tested by **simulating** all types of **accidents**. New models are tested by **experts** from specialty publications who also evaluate other aspects such as the **esthetics**, **comfort**, and the relationship between **quality and price**.

The automobile industry is one of the most important economic sectors, and the automobile has turned into a symbol of our time.

AUTOMOBILE COMPONENTS-BODY AND FRAME

Cars and trucks, like the old-time coaches, are assembled on a **chassis** that consists of a **steel** frame onto which the body and the mechanical components are assembled. Modern cars, on the other hand, use **monocoque** or **unibody construction** that takes the place of a chassis. Some bodywork is made of **aluminum** or **fiberglass**, but most of it is still manufactured from **sheet steel**.

RECYCLABLE MATERIALS

Most of the materials used in manufacturing new automobiles can be used again. There are special plants for that purpose where some components such as tires, plastic parts, and copper wire are separated, and the steel and iron parts crushed for subsequent melting in blast furnaces.

 Tractors, self-contained cranes, and trucks are also automobiles in the true sense of the word.

THE ELECTRICAL SYSTEM

The central component of the electrical system is a 12-volt **battery** that is recharged by a **dynamo** or an **alternator-rectifier** system while the motor is running. In addition to making it possible to start the motor, the battery provides the energy for the **lighting system**, the **gauge** sensors such as the fuel level and temperature, audio equipment, air conditioning, electric doors and windows, and so on. The system is completed by the **wiring** necessary for it to function and the **fuses** that prevent short circuits.

PARTS OF A DYNAMO

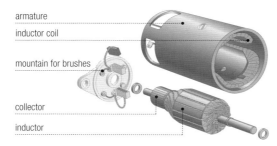

armature
inductor coil
mountain for brushes
collector
inductor

The dynamo is used to recharge the battery.

A HIGH-VOLTAGE SPARK

The compressed mixture of fuel and air that each cylinder contains works like an electric insulator, so a 12-volt spark produced by modern batteries would be incapable of causing the necessary explosion; as a result, automobiles are provided with a **high-tension coil** that boosts the voltage up to 10- or 15,000 volts.

THE TRANSMISSION SYSTEM

The transmission is made up of the mechanisms that make it possible to **apply** the **force** of the motor to the automobile's **forward motion**. It consists of three main parts: the **gearbox**, the **clutch** and the assembly that is composed of the **drive shaft**, the **rear axle**, and the **differential**. The **gearbox** is composed of **gears** of different diameters whose function is to **reduce** or **increase** the number of revolutions that the motor transmits to the wheels. The **clutch** connects or disconnects the motor from the gearbox so it can run while the car is stationary and while the driver is changing gears.

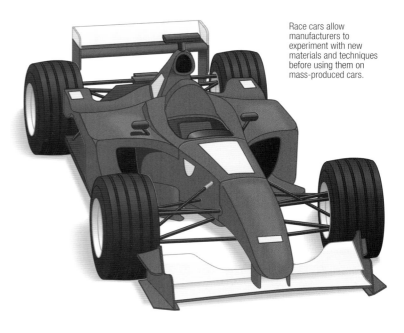

Race cars allow manufacturers to experiment with new materials and techniques before using them on mass-produced cars.

THE DRIVE SHAFT, THE REAR AXLE, AND THE DIFFERENTIAL

The drive shaft transmits the **spinning motion** from the gearbox to the **differential**. It is an axle that spins around and meshes with the gears of the differential, which **change** the rotational direction. The **rear transaxles**, which come out of both sides of the differential, communicate the movement directly to the **wheels**. The whole assembly changes the **direction** of the movement so that it is what is needed for propulsion.

TIRES, SUSPENSION, AND BRAKES

The **shock absorbers** are features designed specifically to **absorb** the vibrations produced by **irregularities** on the road surface and to allow for a slight **lean** in **curves**. The four transaxles end in a part called the **hub**; where the metal wheels are mounted that hold the rubber tires. The **brakes** are located between the hub and the wheel; generally these are metal disks that spin with the wheel and **press** against it when the **brake pedal** is pushed.

Tires, which are mounted on the wheels, are an essential ingredient in the safety and efficiency of a vehicle. There are many types of tires intended for the type of work for which the vehicle is designed.

THE STEERING SYSTEM

The steering system is comprised of the **steering wheel**, the **steering column**, the **connecting rods**, and the **ball joints**. Its function is to transmit to the front wheels the movement that the driver imparts to the steering wheel; **mechanical** and **hydraulic** devices are used in that process.

BUSES AND TRUCKS

CROSS SECTION OF A DIESEL TRACTOR TRAILER TRUCK

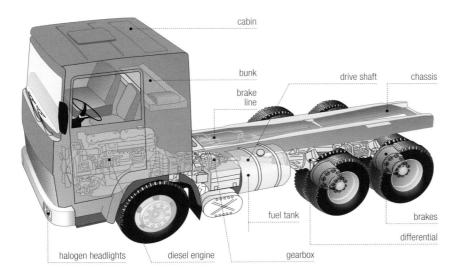

cabin

bunk

brake line

drive shaft

chassis

halogen headlights

diesel engine

fuel tank

gearbox

brakes

differential

Trucks and buses have a mechanical setup similar to that of pleasure vehicles, but they have a few features that set them apart. Most of these vehicles have **extra** braking systems that depend on the **electrical system** or the **motor**, and that increases the safety of heavy vehicles. Many of them have more than two axles; with trucks there can even be more than four, in the case of special trucks for extremely heavy loads. Many modern trucks have a **tractor** where the **cabin** and the mechanical components are located, which is hooked up to a **trailer** for the cargo.

Double-decker buses make it possible to transport many people without taking up lots of room.

MOTORCYCLES

The first motorcycles, which were made at the beginning of the twentieth century, were little more than bicycles with a small motor attached. They have not stopped evolving since that time, and today there are lots of models that differ from one another in appearance and the power of their motors. They are propelled by **two-cycle** engines of one to four cylinders. Their transmission system involves a **gearbox** with no **reverse**, plus a **chain** that communicates the movement from the gearbox to the **rear wheel**. The **steering** is provided by a **handlebar** attached to the front wheel, where the shock absorbers are located; and the **chassis**, which is located between the two wheels, includes all the mechanical elements and the driver's seat.

THE PARTS OF A MOTORCYCLE

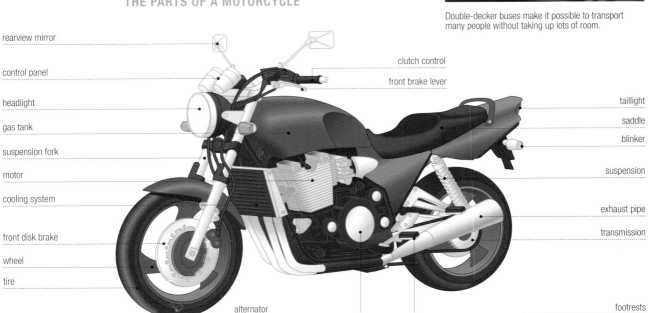

rearview mirror

control panel

headlight

gas tank

suspension fork

motor

cooling system

front disk brake

wheel

tire

alternator

clutch control

front brake lever

taillight

saddle

blinker

suspension

exhaust pipe

transmission

footrests

AIR TRAVEL

The dream of flying like a bird seemed to be among the least likely feats to be achieved, at least until very recent times. Near the end of the eighteenth century a study of the behavior of **gases** made it possible to construct **aerostatic balloons** that used **hot air** to rise; however, it wasn't until a century later, at the end of the nineteenth century, that a way was discovered to equip them with a steering system. **Gliders** made of fabric and light wood also became popular; they used warm **air currents** to take flight, and they were quite difficult to control.

Currently there are several projects underway to reintroduce the dirigible as a way of carrying cargo, especially over moderate distances, because they don't need much room to land.

PRINCIPAL COMPONENTS OF A HOT-AIR BALLOON

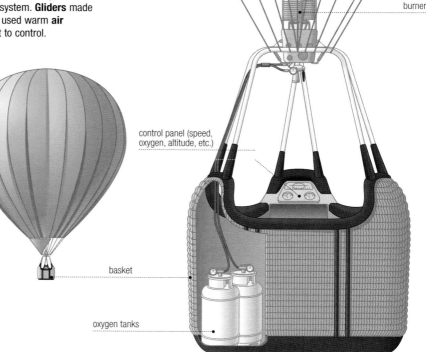

burner

control panel (speed, oxygen, altitude, etc.)

basket

oxygen tanks

THE DIRIGIBLE, THE FIRST SERIOUS ATTEMPT

The Idea of attaching a **steam** engine to move a **propeller** in the carriage of a hot-air balloon **failed** repeatedly, and even when the lighter-weight **internal combustion engine** was invented, it was still impossible for the balloon to overcome the **resistance** of the air. The German inventor August Heinrich von **Zeppelin** built an **aluminum** balloon in a cigar shape to cut down wind resistance; propelled by internal combustion motors, it succeeded in flying on July 2, 1900, ushering in the era of air transport.

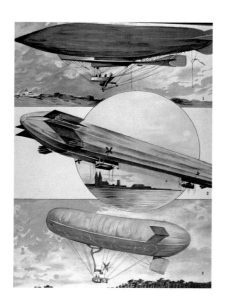

Three types of dirigibles from the beginning of the twentieth century: above, Patrie; middle, Zeppelin; and bottom, Percival.

DOGGED BY TRAGEDY

The **dirigible** balloon or the **zeppelin**, as it was also known in honor of its inventor, showed for almost 30 years that it could be used not only for short and moderate flights, but also for **ocean crossings**, with the added advantage that it did not need specially prepared areas for **takeoff** and **landing**. It was successful as a means of transporting people and goods until a series of accidents capped by a major tragedy caused them to be outlawed. The cause was that the balloon was filled with **hydrogen**, a gas that is much lighter than air but extremely **flammable**. Present-day dirigibles, used especially for advertising purposes, use **helium**, which is much more **expensive** but **harmless**.

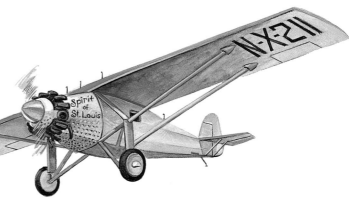

Given the large size and lack of maneuverability of the dirigibles, the airplane caused a real revolution in air travel. The illustration shows *The Spirit of Saint Louis*, Charles Lindbergh's plane. He completed the first transatlantic flight in 1925.

THE FIRST AIRPLANES

Gliders had existed for 50 years when, in 1903, brothers Wilbur and Orville Wright succeeded in getting one of them to fly for 820 feet (250 m) using an internal combustion motor. That was the culmination of many failed experiments by other aviation pioneers over the preceding years. In order to achieve that the brothers had to invent the **movable ailerons** that made it possible to control the craft, and to perfect a strong but light **motor**.

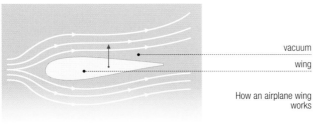

vacuum

wing

How an airplane wing works

MODERN AIRPLANES

There are many types of airplanes today: small **private planes** with one or two seats powered by a **propeller**, large planes that carry more than **300 passengers** in transoceanic flights, propelled by large **jet engines**, planes with **two propeller** engines, or **two or three jet** engines.

CROSS SECTION OF A COMMERCIAL PLANE

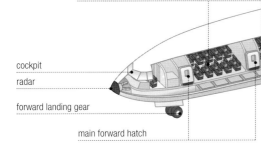

passenger compartments

cockpit
radar
forward landing gear
main forward hatch

landing gear

HOW DO AIRPLANES FLY?

In order for a body that is **heavier than air** to fly, it must be acted upon by a **force** at least equal to its weight and operating in a direction **opposite** the center of the earth. **Birds** that soar, **hang gliders**, **gliders**, and **parasails** get that from rising **currents** of **heated air**, but birds have to beat their wings and the devices mentioned cannot fly by themselves. **Airplanes** take advantage of the fact that air cannot completely flow around objects that move through it at high speed; they use wings whose **shape** creates a **vacuum** on the upper surface so that the **force** exerted by the air **under the wing** compensates for the weight of the plane. This effect depends on both the **speed** of the plane and the **size and shape** of the wing.

It seems unnatural to see such heavy devices suspended in the air, but airplanes have shown themselves to be amazingly safe.

stabilizer
rudder
swept-back wing
vertical stabilizer
ailerons
jet
cargo bay
fuselage

HELICOPTERS

In 1939 the American engineer Igor Ivan **Sikorsky** invented a machine **with no wings** that can fly by means of a large propeller located on its upper surface. The low pressure created by the rotation of the propeller keeps the helicopter up in the air, even stationary, but a second, smaller propeller spinning in a plane perpendicular to the other one is needed to control the aircraft.

Even though helicopters cannot reach great speeds, they are extremely maneuverable.

PHOTOGRAPHY, MOVIES, AND GRAPHIC ARTS

People have been fascinated by the production of images ever since prehistoric times, as demonstrated by the **cave paintings** found in various parts of the world; however, the artistic and small-scale methods used made them available to only a few people. The combination of **printing** and **photography**, on the other hand, and mass distribution of **movies** have put images within reach of everyone.

LIGHT ABSORPTION AND REFLECTION

When an object is illuminated it absorbs part of the light that it receives and reflects the remainder. Since **white light** is the result of **combining** all the wavelengths from **red** to **violet**, which comprise the **visible** spectrum of **electromagnetic waves**, every illuminated surface has its own tendency to reflect some wavelengths, thereby producing the **color** that we perceive. Tree leaves are **green** because the wavelengths that correspond to this color are the only ones that are reflected.

THE PRINCIPLE OF PHOTOGRAPHY

During the eighteenth century it was discovered that some **silver salts became darker** when they were exposed to light, and thereafter, people began experimenting with procedures to render permanent the images thus captured. The idea involved impregnating a smooth surface, such as a piece of glass, with silver salts, protecting it from exposure to light, and then exposing this **photosensitive plate** for a certain time to the light reflected from an object. Since every point on the object reflects a different quantity of light, a **negative** image of the object is formed on the plate. The points that **reflect the most light** are the ones that **darken** the salts the most, while the ones that absorb **all** the light remain **white.**

Various types of photographic cameras.

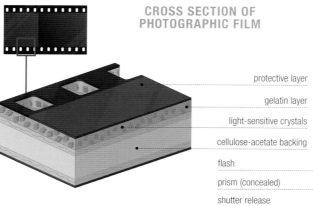

CROSS SECTION OF PHOTOGRAPHIC FILM

protective layer

gelatin layer

light-sensitive crystals

cellulose-acetate backing

There are several types of photographic solutions that are distinguished from one another by their different abilities to use light. Manufacturers use DIN or ASA numbers to identify them. The highest numbers correspond to very sensitive films.

PRIMITIVE CAMERAS

The best results were obtained when light was allowed to pass through only a tiny **aperture** in a **hermetically sealed box** at the back of which was located the photosensitive plate. Subsequently, a **convex lens** was added inside the hole, resulting in a primitive **photographic camera**.

DIAGRAM OF A REFLEX CAMERA

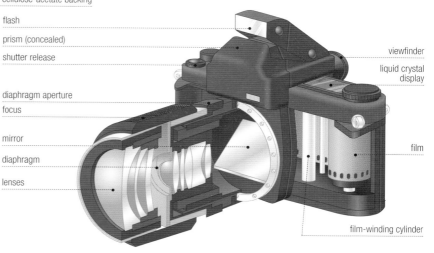

flash

prism (concealed)

shutter release

diaphragm aperture

focus

mirror

diaphragm

lenses

viewfinder

liquid crystal display

film

film-winding cylinder

THE EVOLUTION OF PHOTOGRAPHIC FILM

Once the appropriate silver salts were obtained, a method was developed for dissolving them in a **gelatin** coating on a hard base. At first glass was used, but this was soon replaced by **celluloid**. Films that produce **color** photographs have three separate layers of **red**, **blue**, and **green** sensitive coating. During **developing**, the **red** layer produces **cyan** (deep blue), which is the result of eliminating the red from white light; **magenta** (purplish red) is produced in the **blue** layer; and **yellow** is produced in the **green** one. The colors in the developed photograph are the result of the mixing of these three colors in different proportions.

The most common film is 35 mm, which results in a 24 × 36 mm format.

DEVELOPING PROCESS FOR BLACK-AND-WHITE FILM

1. Enlarger; the photographic paper is exposed to the negative image. This yields a positive image.

2. The paper is submerged in developing solution.

3. Washing the image.

4. Fixing the image

5. Washing the image again

6. Drying

DEVELOPING

To make visible the changes produced in the **film** during the time it was **exposed** to the light, when a photo is taken, it has to be subjected to a process that is carried out in a **darkroom**, an area of the photo lab where work is done in a faint, red light. First the film is immersed in a developing solution that has to be stirred constantly. Once the recommended time has passed, the film is transferred to a **neutralizing** bath that stops the action of the developer, and then it is subjected to the action of a liquid **fixer**. To finish up, it is washed with water for several minutes to eliminate any vestige of the chemicals used, and after it is dried, the **negatives** are ready to be turned into **positives**.

THE POSITIVE IMAGE

In order to convert the **negatives** into **positives**, which are the photographs that we are accustomed to seeing, they have to be put into an **enlarger** that projects the negative image onto the **photosensitive paper** on which the photographic image is to be printed. As with negatives, the result is invisible until the paper is subjected to the same operations as applied to the film. This is how hobbyists develop film, since professional developing is highly automated.

A special light has to be used in the dark room to avoid ruining the photograph.

In 1839 Daguerre used sheets of plated copper that had been exposed to the action of iodine vapors to form a film of silver iodide, which is very sensitive to light. Upon exposure they form a positive image of extraordinary accuracy and beauty. The system, which is known as daguerreotype, was abandoned because of its dangerous nature.

REPRODUCTION

Although it is possible to make any number of prints from a negative, other means are used for very large distribution of photographs. The **printing** techniques used in modern **graphic arts** make it possible to reproduce full-color images much more economically. This is because the modern printers are completely automated and can print thousands of copies in a few hours, and because common paper is much more inexpensive than photographic paper.

PRINTING TECHNIQUES

In the printing technique invented by **Gutenberg** in the fifteenth century, words were formed by placing the letters one by one in guides, and when the page was complete it was coated with ink and pressed onto paper; the operation was repeated as often as necessary to produce the desired copies. The processes for composing texts evolved slowly, and in the nineteenth century they achieved a certain degree of **mechanization**, but the technique that made possible the true modernization and popularization of graphic arts was **photocomposition**.

The printing process is now fully automated, producing thousands of books or magazines in an hour.

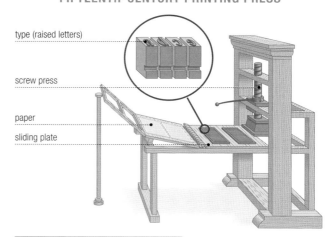

type (raised letters)

screw press

paper

sliding plate

COLOR PRINTING

In color printing four different plates are printed separately and used for black, yellow, cyan, and magenta. The mixing of these colors in different proportions produces all other colors. In rotary presses, the plate that corresponds to each of these colors is mounted on its own cylinder.

PHOTOLITHOGRAPHS

In photocomposition the characters are **photographic** matrices that are placed onto a surface, along with drawings and images, in order to create a photograph known as a **photolithograph**. This is the basis on which the **metal** impression **plates**, which are generally curved, are made; they are placed onto the cylinders of **rotary presses**, the modern printing machines. With these machines, a **continuous** roll of paper passes between the printing **cylinders** that contain the plates and the ink.

In four-color printing, a photolithograph of each of the basic printing colors is required.

yellow

magenta

cyan

black

four-color

One of the most commonly used printing processes is offset, in which the ink that coats the metal plate is transferred to a sheet of rubber rolled onto a cylinder, which in turn transfers it to the paper.

DIGITAL PRINTING

In modern printing systems, the entire process is guided by computer. In place of photolithographs, the printer receives a compact disc that contains both the text and the digitized images. A computer interprets the information contained on the disc and transfers it to the machines that prepare all the printing elements based on the instructions received. Technicians check the results with proofs and input last-minute changes if needed.

MOVING IMAGES

The human eye retains for a few tenths of a second the light images that reach it, in such a way that if the **movement** of a figure **is broken down** into several **successive static images** that are run before the eyes at a speed of at least 12 images per second, the human brain perceives them as **images in motion**. This phenomenon, which has been known for a very long time, was applied during the nineteenth century to devices with such complex names as **phenaskistiscope**, **praxinoscope**, and **fantascope**, the most familiar example of which is the **zootrope**. This consisted of a series of illustrations placed on a cylinder that spun quickly and caused them to pass in front of a slot through which the observer watched, making the images appear to move.

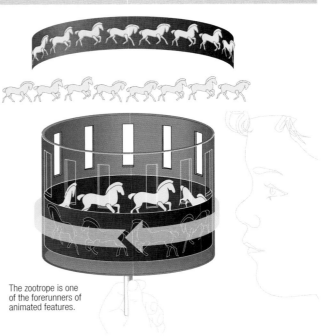

The zootrope is one of the forerunners of animated features.

Special effects make it possible to create spectacular and unreal scenes. The image shows rigging for the movie *Batman II* in Hollywood studios.

CINEMATOGRAPHERS

Although other inventors made important advances in projecting moving photographic images, in 1895 the brothers Auguste and Louis **Lumière** were the first to present a public **projection** of a film shot and projected using equipment that they had developed. Silent films were projected at a rate of **16 images** a second, but when **sound** was introduced, film was projected at **24** images per second.

FILMING

When the images of a movie are filmed, the tape is kept in front of the lens with the shutter open for **1/48th** of a second; then the shutter is closed and the tape **advances** with the shutter closed for the same amount of time, and so on. Cinematographic film is developed using processes similar to those of conventional photography, but professional film has to undergo some further processing before being shown. During **assembly** the tape is cut and spliced so that the images on film appear in the order desired by the director; this is also where the **touching up**, **special effects**, and the **sound** are added. If sound is used, it is recorded separately and added to the **sound track** that accompanies the images on the film.

↓

Photography, movies, and television quickly became indispensable features of modern society.

MAIN COMPONENTS OF A MOVIE CAMERA

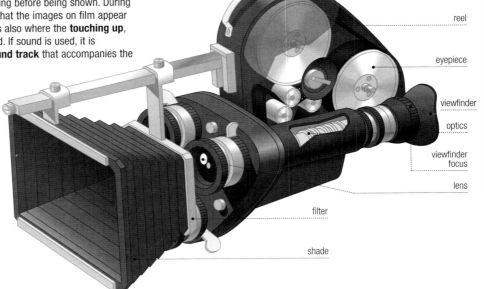

reel

eyepiece

viewfinder

optics

viewfinder focus

lens

filter

shade

ELECTRONICS

Electronics arose as a division of electricity when **Edison** accidentally discovered that an electrical current flows from a hot filament. This was later identified as coming from **free electrons** when in the presence of another current charged with positive electricity. This phenomenon was used to develop such familiar devices as radio, television, computers, stationary and mobile telephones, and reproducers of images and sound, plus such necessary ones as radar and sonar.

THE BIRTH OF ELECTRONICS

People began using electricity a long time before the causes of electric phenomena were understood. In 1897 the British physicist Joseph J. **Thompson** identified the **electron** as a particle charged with **negative** electricity, and in 1898 the German engineer **Wein** discovered that the **proton** is responsible for positive electrical charges. Up to that time, scientists did not understand the nature of electricity. These discoveries provided a theoretical basis for the discoveries of Edison and of **cathode rays** by the German Johan W. **Hittorf** in 1869.

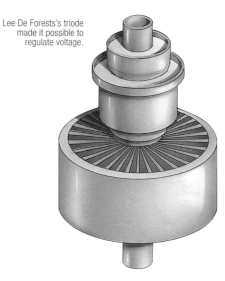

There is a very sophisticated technology behind a familiar gesture such as speaking on a cordless phone.

THE FIRST DIODE

The physicist Ambrose **Fleming** studied the **Edison effect**, and in 1904 invented a device comprised of two **electrodes** located inside a bulb from which the air had been removed. When electricity was introduced to it, the **cathode**, or negative electrode, heated up and emitted an electron current that flowed toward the **anode**, or positive electrode. The latter consisted of a metal plate so that the circulation of electrons took place in a single direction. That made is possible, among other things, to convert **alternating** current to **direct**.

DE FOREST'S VALVE

In 1907 the American engineer Lee **De Forest** perfected the **diode**, or Fleming's **lamp**, by adding a fine **metal grate** between the **cathode** and the **plate** connected to the electric current. It functioned in such a way that when an electrical current flowed through it in a direction opposite that of the **cathode**, **fewer** electrons reached the **plate**. By regulating the **voltage**, it functioned as a real **valve** that controlled the electron flow. Since it is composed of three parts, it is also known as a **triode**. Both this device and the diode had many applications in the recently invented **radio**.

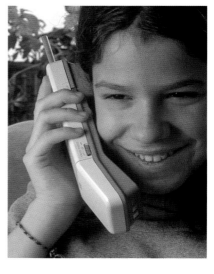

Lee De Forests's triode made it possible to regulate voltage.

Fleming's diode made it possible to convert alternating current to direct.

Rapid developments in electricity starting in the middle years of the twentieth century have revolutionized the lifestyles of hundreds of millions of people, especially in developed countries.

TRANSISTORS

In 1948 the Americans **Bardeen**, **Brattain**, and **Shockley** invented the **transistor**, an electronic device that would come to revolutionize this technology. Transistors take advantage of the fact that the outermost electrons of atoms tend to detach, attract other electrons, or form bonds with their counterparts in other atoms. In the solids of **column IV** of the periodic table, which includes **silicon**, the outer electrons form bonds and crystals that **don't conduct electricity**, but if an atom from an element of **group V** replaces an original one, there is one **free electron** that can move through the crystal, and if the atoms that are introduced are from **column III**, a **positive charge** is liberated.

CROSS SECTION OF A BIPOLAR TRANSISTOR

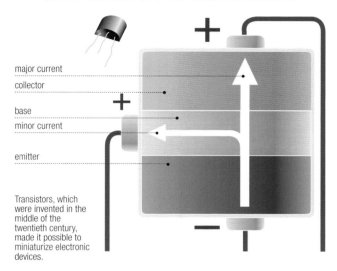

major current
collector
base
minor current
emitter

Transistors, which were invented in the middle of the twentieth century, made it possible to miniaturize electronic devices.

HOW TRANSISTORS WORK

When the atom introduced is from group V, such as **phosphorus**, it creates a **semiconductor crystal** of the **n** type; but if it is from group III, such as **boron**, it will be of the **p** type.

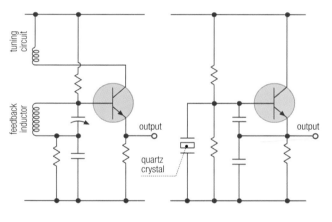

tuning circuit

feedback inductor

output

quartz crystal

output

On the left, a diagram of a synchronized oscillator; on the right, a diagram of a piezoelectric oscillator.

TYPES OF TRANSISTORS

There is a great **variety** of different transistors that perform **electrical functions**, but they all are based on the different possible combinations involving layers of semiconductor crystals of the p and n types. So, for example, a transistor of the **n-p-n** type, in other words, which has **one positive** semiconductor crystal between **two negative** semiconductors, can serve as an **amplifier** of current, depending on how the crystals are connected to it.

These bipolar transistors serve to amplify the current.

OTHER ELECTRONIC FUNCTIONS

In addition to **rectifying** current, such as n-p, or **amplifying it**, as with n-p-n**,** transistors are also used for electronic functions such as oscillation, which involves the rhythmic alternation in current intake necessary for **radio receivers** and **electronic watches**, **commutation**, in which they are designed to **interrupt** or allow the passage of an electric current at will, or for **timing**, when they are used to allow the current to pass for a time and cut it off at the end.

Since transistors can be manufactured very economically, they have made possible many electronic devices available to millions of middle-class families all over the world.

CIRCUITS

An electronic circuit is formed when there is at least one **electric lamp** or a **transistor** between the current input and output. Therefore, there can be millions of different electric circuits by taking advantage of the possible combinations among various elements, but in practice, the useful combinations amount to only a few hundred.

CHIPS

The word **chip** is sometimes used to designate French fries, and since the slender **silicon wafers** that are used to manufacture **integrated circuits** resemble them, this term has been used popularly to describe the wafers. In each one there may be from just a few to thousands of **valves**, **transistors**, **resistors**, and **condensers**. Since most of them are very complicated, before manufacturing them technicians design a map with all the components that make up the chip; in many cases this has to be done with the aid of a computer.

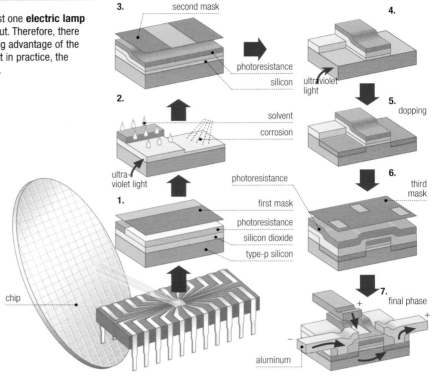

3. second mask
photoresistance
silicon
4.
ultraviolet light
5. dopping
2. solvent
corrosion
ultra-violet light
photoresistance
1. first mask
photoresistance
silicon dioxide
type-p silicon
6. third mask
7. final phase
aluminum
chip

MANUFACTURING CHIPS

The raw material used for manufacturing chips is type **p** silicon, which normally is obtained from **quartz crystals** cut into a cylindrical shape so that very thin **slices** with a surface area of about a half-square-inch (1 cm) can be sliced off. The design that the technicians have done previously is **printed** onto the chips with a photographic process by **covering** certain areas and leaving others **exposed** to the actions of **corrosive acids**. The process is repeated until the different **layers** of the chip are created. The entire process is carried out under the strictest standards of cleanliness, since even the tiniest speck of dust can render the chip useless.

MICROCHIPS

The miniaturization process has continued, and today, companies manufacture **microchips** that contain up to **nine million** transistors with no increase in size, and it is predicted that this astronomical number may double.

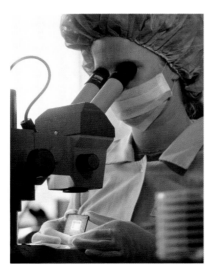

Microscopes are needed for quality control with chips because of their diminutive size.

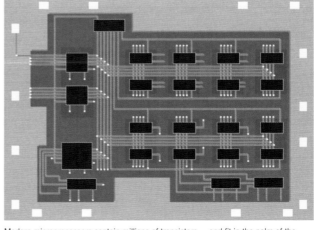

Modern microprocessors contain millions of transistors ... and fit in the palm of the hand.

RADIO

The radio is a contemporary of the electronics industry and has benefited from electronics advances. Its main characteristic is that it uses **electromagnetic waves** to **carry signals** of different types, including **sound**, instead of using electrical **cables**, as the **telephone** and **telegraph** have done. To accomplish that, in 1888 the German physicist Heinrich **Hertz** discovered a way to produce long-length electromagnetic waves, and a few years later the Frenchman **Branly** developed a rudimentary **receiver** for these waves. Then the Englishman **Lodge**, the Russian **Popov**, and the Italian **Marconi** discovered the usefulness of **antennas**, with the result that Marconi was able to conduct the first **radio** broadcast in 1901.

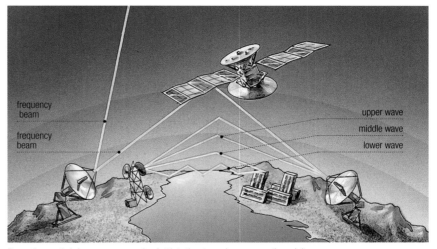

frequency beam

frequency beam

upper wave
middle wave
lower wave

Depending on their frequency, waves ricochet into the atmosphere or pass through it.

WAVE MODULATION

In order to produce radio waves, **electrical and magnetic** fields perpendicular to one another are made to vibrate, but in order for these waves to carry sound they have to be modified by **modulating** one of its two basic properties, the **amplitude** or the **frequency**. AM stations broadcast by modulating the amplitude, and **FM** stations modulate the frequency. The **frequency** is the number of times that a wave **vibrates** in every second; the **amplitude** is the **length** of a complete wave.

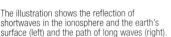

ionosphere
shortwaves
Earth

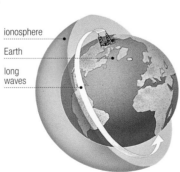

ionosphere
Earth
long waves

The illustration shows the reflection of shortwaves in the ionosphere and the earth's surface (left) and the path of long waves (right).

CARRYING THE SIGNAL

The modulated and amplified wave is sent into the air from an **antenna**, and it **expands** in all directions. First, however, the **sound** of the voice or the music has to be converted into electrical impulses, a task that is carried out by **microphones**. These have a membrane that vibrates in response to sound waves, producing an **electromagnetic field** and creating the impulses that subsequently are amplified and broadcast. A **receiver** connected to an **antenna** is needed to **receive** the waves; it must be **tuned** to the same amplitude or frequency as the broadcasting unit.

USING WAVES

The frequency of radio waves varies between 3 and 3×10^7 **kilohertz**. As a result they have been divided into different **frequency bands** that are used for such purposes as broadcasting time signals, navigation signals, and commercial radio and television.

CONVEYING SOUND WAVES

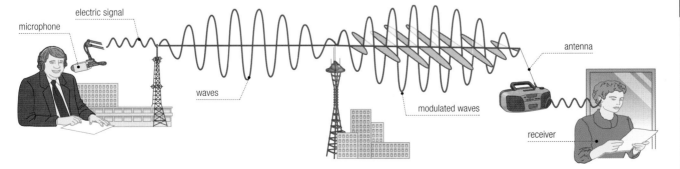

electric signal

microphone

waves

modulated waves

antenna

receiver

TELEVISION

As with movies, the **sensation** of **movement** in televised **images** depends on the capacity of the human eye to retain them for a few tenths of a second after the stimulus from the light has ended. However, the way of **producing**, **analyzing**, and **transmitting television images** is based on principles entirely different from the ones used in the movies. In 1884 the German **Nipkow** developed a spinning **disk** with small **openings** distributed in a **spiral** pattern. As it spun, the disk **broke down** the images and created **lines** that started at the outermost opening and descended to create **parallels** using the inner holes.

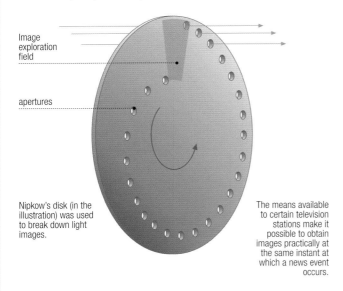

Image exploration field

apertures

Nipkow's disk (in the illustration) was used to break down light images.

The means available to certain television stations make it possible to obtain images practically at the same instant at which a news event occurs.

MODERN CAMERAS

Today, television cameras use very different methods from the one represented by Nipkow's disk to analyze and break down images. An **exploration point** goes along the lines and breaks down every image into more than 300,000 basic data involving color, contrast, brilliance, and depth, which amounts to some 4,000,000 data per second. **Digital cameras** convert these data into **numbers**, which are easier to process and store.

BREAKING DOWN THE IMAGES

Nipkow's disk took advantage of the property of **silenium**, which was discovered in 1873, whereby its ability to **conduct electricity** varies according to the **amount of light** that it receives. The light that was allowed to pass through each aperture of the disk worked on a **photoelectric cell** that converted the light values into **electric impulses**. This was the signal that was **amplified** for broadcasting through an **antenna**, like radio waves.

IMAGE TRANSMISSION

When television transmission is carried out through waves, the **very high frequency** (**VHF**) or **ultra high frequency** (**UHF**) bands are used; they have the necessary amplitude to convey such a tremendous quantity of information. The television signal is transmitted by modulating the amplitude of the carrying waves (**AM**), which have to be amplified several times before being broadcast through the antenna. This type of transmitting television signals limits the number of channels that can operate in the same area, and that is why it is increasingly common to use **fiber-optic cable** to conduct the signals.

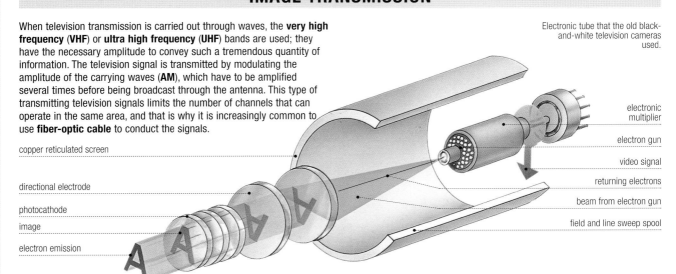

copper reticulated screen

directional electrode

photocathode

image

electron emission

Electronic tube that the old black-and-white television cameras used.

electronic multiplier

electron gun

video signal

returning electrons

beam from electron gun

field and line sweep spool

RECEPTION

The **antenna** receives the modulated wave and converts it into weak electrical impulses that reach the **tuner** on the receiver, which separates them from other currents of different frequency. After being **amplified** several times, the signal reaches the **picture tube**, which emits **electron beams** that sweep it and produce the same **image** that the cameras of the television station to which the set is tuned captured a fraction of a second earlier.

RECEIVER TUBES

The most important part of a television set is the funnel-shaped **vacuum tube** that ends in a flat plane that forms the **screen**. Inside the tube there are three **electron guns** surrounded by an electromagnetic spiral that receive separate signals that correspond to the colors **red**, **green**, and **blue**. The electrical impulses that correspond to the signal of each color act on these spirals, whose function is to **divert** vertically and horizontally the beams from the electron guns, so that they reconstruct on the screen the image being broadcast by the station.

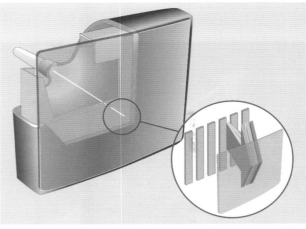

In the Trinitron system, the color beams pass through the mask separately, and to create greater contrast in the picture, it is strengthened with black luminophores.

DIAGRAM OF A COLOR TELEVISION RECEIVER

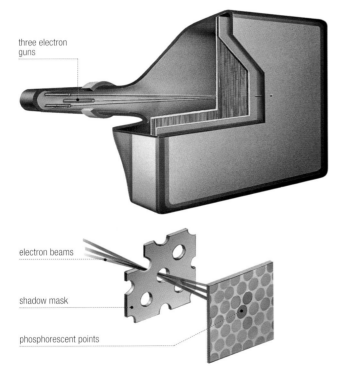

three electron guns

electron beams

shadow mask

phosphorescent points

Currently there are processes that make it possible to receive televised images on flat, plasma, and liquid crystal screens, which may become affordable to the middle-class consumer when manufacturing costs are reduced.

SCREENS

There are several ways to manufacture television screens, but one of the very common ones is to coat the screen with three **fluorescent** compounds that produce the colors red, green, and blue when they are hit by an electron. To protect this sensitive surface, a **black** mask with around 200,000 tiny holes is installed. The three electron beams continually sweep the screen, but each **point** is energized only by the beam of the appropriate color, because it is the only one that reaches it at the **proper angle**.

The annoying flickering in the screen has been eliminated from modern televisions.

FLICKERING SCREENS

When we see a scene of a movie or a television program on a television set, we see clearly that the image is flickering. In order to make this effect invisible, television stations broadcast twice as many images per second as necessary to produce the sensation of movement.

RADAR

This **electronic detection system** uses **radio waves**; at first it was a military invention developed by the allies during the World War II, but right after that it was adopted by **civilian airports**, which were beginning an extraordinary **expansion** phase. From radar's initial role as a **supplementary safety system**, in a few decades it was transformed into an essential feature of modern **commercial aviation**.

Mobile radar used during World War II. Radar radically changed defense systems.

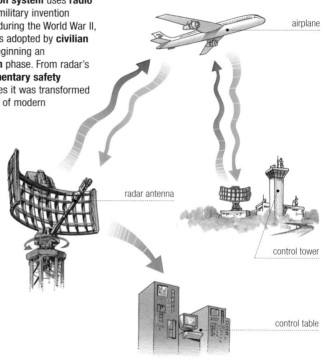

airplane

radar antenna

control tower

control table

HOW RADAR WORKS

Radio waves between 1,000 and 35,000 **megaherz** are sent out from the radar installation. They travel around 180 miles (300 kilometers) a second, and when they strike a solid object they bounce back to their place of origin, where an electronic system analyzes this **echo** and calculates the **time** elapsed since the signal was first sent. In order to avoid confusing echoes from the signals, it is possible to send **short and intermittent** signals so that no new signal is sent until the **echo is received**; but when the signals are sent continuously, the impulses are **rhythmical**, increasing and decreasing the signal's **intensity**. The data collected are displayed on a cathode ray **screen**.

Radar has made it possible to control air traffic with total safety. The radar antenna pinpoints the precise location of each airplane and displays it in the control room so that the air traffic controllers can give appropriate instructions to the pilots.

SONAR

Sonar is a process used to **detect** the presence of objects under water, using **sonar** waves. The sonar system can be either **active** or **passive**, depending on whether it **sends** sounds or merely **receives** sounds from some other source. The most important **civilian** uses for sonar include locating **schools of fish** and **mapping out** the ocean floor. This involves using a sonar system exterior to the boat to explore the bottom with a beam that is very narrow horizontally but broad vertically.

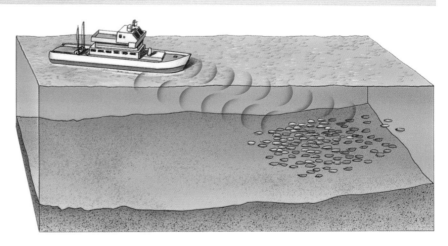

Sonar installed on ships makes it possible to detect objects underwater such as a school of fish, a sunken ship, rocks, etc.

Presently, modern fishing vessels in industrialized countries depend on sonar; that has increased their take significantly, but it has also endangered many marine species.

HOW SONAR WORKS

The sonar signal is sent from an **underwater transducer**, a device that converts **electrical** impulses to **sonar** waves. Since sound travels in water at a speed of just under 1 mile (1.5 km) a second, the **distance** from the transducer to the object reflecting the sound can be calculated based on the time elapsed until the **echo** is received. Since the intensity of the sound **decreases** as **distance increases**, the effective range of sonar varies between 100 yards/meters and 6 miles (10 km).

X RAYS

Taking pictures through solid objects seemed to be an unrealistic prospect until X rays were discovered. They work by imprinting a photographic plate with varying shades of gray depending on the nature of the solid masses through which the rays pass. These rays are named after the German Wilhelm Konrad **Röntgen**, who was unfamiliar with the nature of the **radiation** that he had discovered. In the early decades of the twentieth century, he discovered that these were **electromagnetic waves** with a shorter wavelength than light, in a broad band that included radiation of different **penetrating power** and varying degrees of **danger**.

The first X ray in history shows the hand of Berta, W. K. Röntgen's wife.

RADIOLOGY

Radiology is the science concerned with the application of X rays to both medicine and to industry. This was the means by which doctors first obtained images of the **inside** of the human body—initially the **hard** parts such as **bones** and **cartilage**, which made it possible for specialists to get a detailed view of **fractures**, torn **ligaments**, and **deformities** in any of these features, and subsequently the soft organs such as the stomach and liver. In order to take X rays of these organs it became necessary to develop substances that could be ingested to serve as a **contrast** in the particular organ; these had to be easily **assimilable** and constitute no health hazard.

INDUSTRIAL RADIOLOGY

In the medical field efforts were centered on developing **contrast** substances and on **security** systems to protect the staff who operated the machines, but in i**ndustry** new procedures were developed that made it possible to **X-ray metal parts** in layers to detect **cracks** or **structural** defects. Radiology became an indispensable tool in **quality control** for parts that are subjected to **great stress** in **precision** devices and in the **aeronautics** industry.

OTHER WAYS TO SEE WHAT IS HIDDEN

Today, procedures other than X rays are used to see inside materials. **Ultrasound** is used both in medicine and industry to create images of the inside of the human body and parts and instruments. Ultrasound images are formed on a screen once they are decoded and interpreted electronically by echo analysis or ultrasound wave reflection from the device.

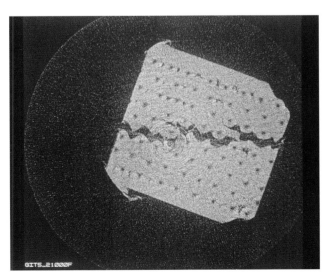

GITS_2100OF

Industrial microscopic radiography makes it possible to study precisely possible fractures in critical materials. The photo show a 50-micron break in a part intended for aeronautics.

specialist

monitor

transducer

transducer

sent ultrasound waves

reflected ultrasound waves

digital system that changes reflected signals into images

image displayed on monitor

Ultrasound make is possible to see such things as the development of a fetus in the mother's womb.

COMPUTER SCIENCE

This new science focuses on **techniques** for automatic information processing. In a broad sense, the purpose of computer science involves **organizing**, **storing**, **transforming**, and **communicating** the information necessary for carrying out any human activity with the aid of a powerful tool, the **computer**.

DATA PROCESSING SYSTEMS

A computer system is the series of interrelated **means** and **procedures** that make it possible to process and communicate information; in other words, it is the set of devices and the method of using them that make it possible to work with information and share it. **Processing** the information involves all the operations that make it possible to convert isolated data into the desired type of information, for example, inputting all the data on traffic accidents during a certain period of time to find out what age groups are most affected, the most frequent causes, and the types of corrective measures that might be used.

HOW A DATA PROCESSING SYSTEM WORKS

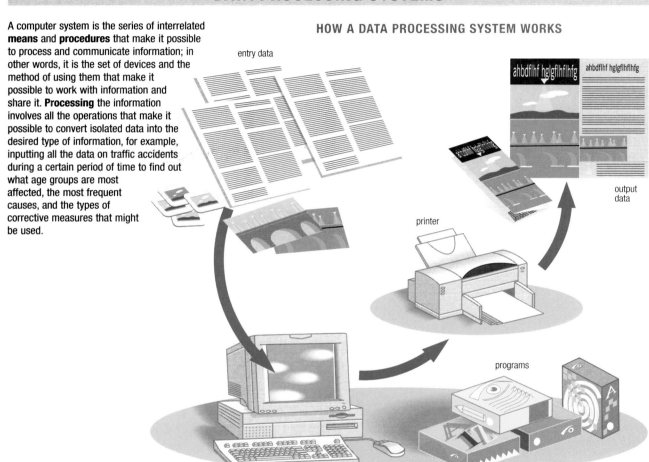

entry data

ahbdflhf hglgflhflhfg

output data

printer

programs

DIAGRAM SHOWING DATA PROCESSING

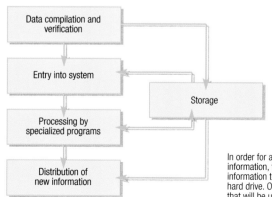

Data compilation and verification

Entry into system

Storage

Processing by specialized programs

Distribution of new information

DATA PROCESSING

Data processing can be broken down into four kinds of **basic operations**: data collection and **entry**, **processing** and **output** and circulation of the reports prepared, **changes** in the presentation of some data, and **choosing** the means to circulate them.

In order for automatic data processing to produce the desired results and produce new, useful information, the steps in the diagram are followed. First it is necessary to compile and verify the information to be processed so it can be stored on diskettes, compact discs, or on the computer's hard drive. Once the data have been compiled and stored, the information is fed into the program that will be used to process it, and once it has been processed, the new information is stored for subsequent processing or distributed by various means.

SUBSYSTEMS

When computer systems are very complex, they are divided into simpler subsystems. For example, the computer system for a hospital covers such diverse fields as the patients' clinical history, data banks for information on the cleaning staff, and programs that run automatic machines and accounting. All these areas, plus many others, have their own computing subsystems.

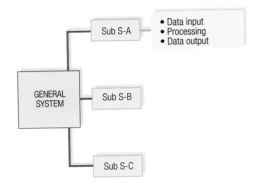

```
C:\>DIR/P

The volume of the C unit is AXIS
The value serial number is 2E1F-11EE
C Directory

.             <DIR>              30/08/98      10:52
..            <DIR>              30/08/98      10:52
SYSTEM   <DIR>                   30/08/98      10:57
COMMAND <DIR>                    30/08/98      10:57
MODEMDET  TXT         58         14/03/99      18:14
NETDET    INI      7.885         24/08/96      11:11
SMARTDRV  EXE     44.867         24/08/96      11:11
HIMEM     SYS     33.367         24/08/96      11:11
REGEDIT   EXE    109.568         24/08/96      11:11
CALC      EXE     59.904         24/08/96      11:11
NOTEPAD   EXE     35.328         24/08/96      11:11
PBRUSH    EXE      4.608         24/08/96      11:11
WRITE     EXE      5.120         24/08/96      11:11
SENDTO   <DIR>                   30/08/98      11:18
SYSTEM    INI      2.129         16/01/01      10:21
Press any key to continue . . .
```

Screen showing a list of programs.

PROGRAMS

In order for these powerful computers to carry out the functions that data processing requires, they have to have all the necessary **instructions** in their **memory**. Computers are machines that have a tremendous **capacity** and **speed** for performing **calculations**, but they need to be instructed in detail how to carry them out. This is the job of computer programmers and amounts to nothing more than a very long list of written instructions in a **language** that the machine is capable of understanding.

Some of the most recent operating systems.

OPERATING SYSTEMS

The basic program that every computer must have is known as the **operating system**. It contains the instructions to make the machine work. The first thing that an operating system does is **identify** the machine where it is to work, since it needs to know the memory it possesses, the speed at which it can perform the calculations, and what means can be used to take in information from the outside. The operating system also **controls** access to any other **programs** that are installed, the **hard disk**, CD or CD ROM readers, and available memory. Some of the best-known operating systems are **MS-DOS**, **Windows**, **Mac-OS**, and **Linux**.

WORD PROCESSING

Word processing programs make it possible to control all aspects of preparing a text such as the **size** and **layout** of the page, the type and size of the **font**, **spaces** between lines, and document **spell check**. The most professional programs make it possible to insert **graphics**, **drawings and photographs**, and **columns**.

DATABASES

Databases are programs for working comfortably with a great number of related elements. They make it possible to select the data by **groups** according to various criteria, and prepare pictures and statistics; however, they require a great deal of **work** to prepare them and constant updating.

The Access relational database.

COMPUTERS

A computer is a machine that is capable of **receiving** information, **searching** through it, **organizing** it, performing **arithmetic** and **logical calculations**, and **writing** the data that it has, with the aid of the necessary programs. In order to do that, the machine uses a **console** or box that contains the electronic devices and to which the **peripherals** are connected. These include the **keyboard**, the **monitor**, and the **printer**, which are essential for receiving and outputting information.

THE FIRST COMPUTERS

The **abacus**, which is still in use, may be the first instrument that humans used for **calculating**, since it was used in **ancient Egypt**, **Mesopotamia**, and **classical Greece**. However, the first **mechanical calculator** was not invented until the seventeenth century, when the Frenchman Blaise Pascal constructed one about the size of a shoebox that was able to add and subtract. Some decades later the mathematician Gottfried **Leibniz** improved it by making it able to multiply and divide; but the first **computer** that was totally electromechanical was developed for **IBM** by the American Howard **Aiken** between 1939 and 1944.

The abacus, the first calculating "machine."

A NEARLY FORGOTTEN APPLICATION

Even though Aiken's machine was the model for the subsequent development of computers, it was not the first one built with those characteristics. In 1938 the German Konrad Zuse introduced the Z1, which had all the characteristics of what we now recognize as a digital computer. This prototype gave rise to the Z2, Z3, and Z4 models, whose improvements made it possible to use them in designing airplanes during World War II, during which they were destroyed. That may be why they are often forgotten in the history of computers.

Home and office calculators are small computers that have a great capacity for performing calculations.

FROM THE MARK I TO THE PC

Computers are an indispensable tool in classrooms.

The first computer, which was named the **Mark I**, needed millions of feet of wiring to function, and it was not **programmable**; in other words, it could perform only the function for which it was built. The **ENIAC** shared the same problem; it was a hundred times quicker than the Mark I and was 98 feet (30 m) long. In 1938 the German engineer Konrad **Zuse** constructed a machine that was programmable, but the first one that used **electronic** elements instead of electromechanical ones was the **ABC**, which was developed by John **Atanasoff** and Clifford **Berry** and served as the model for the computers developed in the **1950s**. With the introduction of the **transistor**, the computers of the 1960s became faster and much smaller, and were the source of today's personal computers.

PRESENT-DAY COMPUTERS

The key to the development of computers to its present appearance and capacity was the **microprocessor**, which makes very small machines **thousands** of times more **powerful** than their predecessors of ten or fifteen years before. The most recent **home** computers are capable of performing more than **a million** simple operations in a single second, and they have enough memory to store several times more information than a large encyclopedia. In addition, there are special computers that multiply by thousands the capacity of these personal computers.

The brain of a computer is known as the CPU, the central processing unit. It includes such features as the control unit, the memory, and the bus.

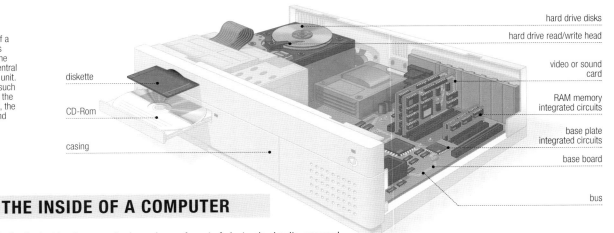

diskette
CD-Rom
casing

hard drive disks
hard drive read/write head
video or sound card
RAM memory integrated circuits
base plate integrated circuits
base board
bus

THE INSIDE OF A COMPUTER

Schematically, the inside of a computer is made up of a set of electronic circuits arranged mainly on the **base board** or the **motherboard**, which is also fitted with **expansion ports** onto which electronic **cards** can be installed for special functions such as controlling **sound** and **video**. The **central processing unit** or CPU is the one that does the **calculations** and controls the movement of data inside the computer. The **hard drive** is the area where the main memory is located, but there are also **memory boards** that can be added to the motherboard.

THE MAIN COMPONENTS OF A DESKTOP COMPUTER

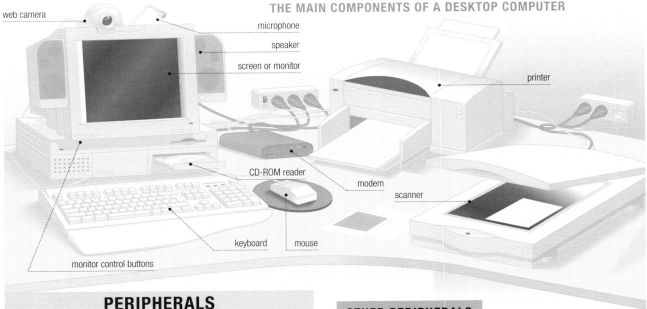

web camera
microphone
speaker
screen or monitor
printer
CD-ROM reader
modem
scanner
keyboard
mouse
monitor control buttons

PERIPHERALS

These are all the elements that make **data entry** and **output** possible. The **keyboard** and **mouse** are the main elements that make it possible to communicate directly with the machine, but there are others as well, such as **joysticks** and **bar code readers**, which have specific uses. **Disk drives** make it possible to input data contained on **diskettes** and **CDs**, as well as copying information contained in the main memory. The set of indispensable peripherals is completed by the **monitor**, on which the data appear.

OTHER PERIPHERALS

Modern computers allow the use of new peripherals to perform new functions; the main one is the **printer**, which transfers to paper the information that appears on the screen. **Speakers** and **microphones** make it possible to hear and record sounds, and **web cameras** gather images that can be sent to other computers.

COMPUTER NETWORKS

Personal computers can be linked together to form **networks** and **share** information. This method of working has been especially useful in medium-sized businesses in a period of growth. Some large companies have used another system that consists of many **terminals** connected to one large central computer, but that way of connecting does not make it possible for the terminals to be linked to one another.

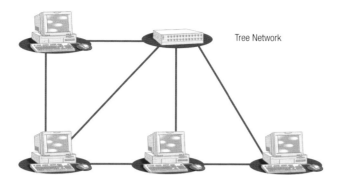

Tree Network

Bus Network

TYPES OF NETWORKS

The connection between a central computer and its terminals is known as a **star network**. When the terminals can also be connected to one another, the network is known as a **tree**. When there is no main computer, the connections can be in a **ring**, in which each machine is connected to the preceding and the following ones, and the last one with the first. They can also be connected in **bus**, when the computers are linked to a single **central channel** of communication and the messages have to travel the whole channel until the computer for which they are intended picks them up.

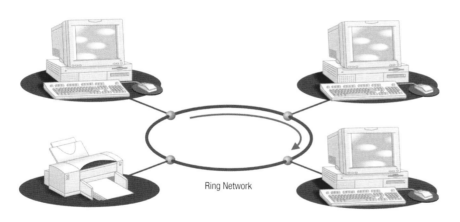

Ring Network

The Internet has developed only a part of its potential. When the financial, commercial, and technical problems inherent to anything so innovative are solved, it will be possible to realize the tremendous advantages that such a powerful medium for accessing information represents.

THE INTERNET

The network that served as the basis for developing the Internet was a military network known as **ARPAnet**; the American army wanted to use it to avoid interruptions to the communications between its units. The idea was that the **network** would decide the **path** that each message would follow to its destination without becoming known beforehand, with the advantage that the **greater the number of computers** connected in the network, the **more possible connection routes**. Some universities and scientific institutions used the same procedure for sharing information, creating a **protocol**, or a series of fixed instructions that all computers could understand.

CONNECTING TO THE INTERNET

The network grew quickly and was turned into a **network of networks** connecting millions of computers throughout the world. The first requirement for connecting to the Internet is owning a communications line, which may be a normal telephone line that requires a **modem** installed in the computer, RDSI or ADSL lines designed to allow faster connections, or a **fiber-optic** line.

The Internet came about as the result of America's military needs.

METHODS FOR CONNECTING TO THE INTERNET

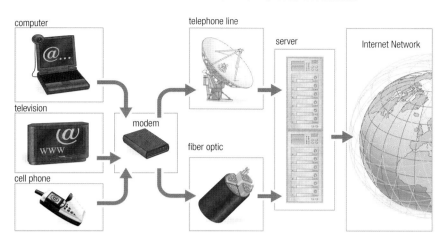

computer

television

cell phone

modem

telephone line

fiber optic

server

Internet Network

Some of the most common activities enjoyed by young Internet users include "chats," the communications that a group of Internet users hold directly and simultaneously on some subject, e-mail, and on-line games in which players from all around the world can take part.

NAVIGATORS AND SERVERS

Navigators are programs created specifically for facilitating Internet connections. **Netscape** and **Explorer** are among the most popular ones, but in addition to owning these programs, users must contract with a service provider or **server** that identifies the user and provides keys to access the network. Many public institutions offer this service free.

WHAT IS THE INTERNET USED FOR?

From the viewpoint of laypeople, the Internet offers two main possibilities: **consulting** an infinity of **web pages**, and **sending** and **receiving** messages. Web pages are documents that contain text and images created in a language that the **navigators** can understand. They are placed in the Web by many persons, companies, and institutions for **consultation** or copying by anyone. The most complete ones contain words or lines that make it possible to access other divisions of the same page or **other pages**, and some of them include **programs** or **games** that can be downloaded onto our computers.

Various web pages.

E-MAIL

Generally, servers offer subscribers the possibility of opening e-mail accounts. They make it possible to communicate directly with persons and institutions whose e-mail addresses are known to us. An e-mail address contains **letters**, **signs**, and **numbers** that identify each user individually. In addition to the address, in order to send or receive e-mail messages, it is necessary to use a **password** that only the user knows and that guarantees **confidentiality** in the communications. The main advantages of e-mail are **speed** and low cost, for messages can be sent to any part of the world for only the cost of a local phone call.

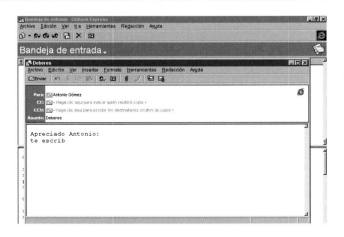

E-mail input screen.

ROBOTICS

Although television and movies have conditioned us to think of robots as mechanical beings that appear human, any machine that can perform a task automatically without the direct intervention of a human can be considered a robot. Scientists hope to construct machines capable of **artificial intelligence** and capable of both performing preprogrammed tasks and learning through experience and making decisions.

COMPUTER-CONTROLLED MACHINES

In the 1970s and 1980s many **computer-assisted** or **computer-controlled** machines were developed, such as **lathes** and **milling machines**. They are capable of executing long and complicated programs, **warn** of possible **accidents** and **breakdowns**, and **correct** errors and faulty adjustments. In the same period there appeared huge **rotary presses** that automated the **printing process** by automatically controlling the quality of the color in illustrations, the ink levels in the tanks, and paper adjustments. During the 1990s **industrial robots** became popular.

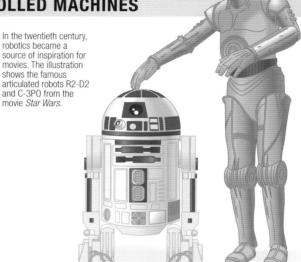

In the twentieth century, robotics became a source of inspiration for movies. The illustration shows the famous articulated robots R2-D2 and C-3PO from the movie *Star Wars*.

Robots can carry out tasks in surroundings that are dangerous for people (as in nuclear power plants) or that require sterile air (as in the pharmaceuticals industry).

ROBOTS IN INDUSTRY

Industrial robots are more than just simple automatic machines, because instead of being designed to perform a single program, they possess enough flexibility to carry out multiple tasks. They have a built-in computer, various sensors for self-orientation, and joints for performing a wide range of movements. Many automobile factories use these robots in different types of operations.

SENSORS

Electronic **miniaturization** and the development of such technologies as **lasers** and **infrared rays** have made it possible to endow robots with **sensors** that come close to duplicating the abilities of the human senses of **sight**, **hearing**, and **touch**. The information gathered by the sensors is processed in the robot's computer, which also makes appropriate decisions for the task that it is performing.

At first the introduction of automatic machines and robots seems to represent a loss of employment for humans; however, as time passes, the machines create more satisfying and higher-paying jobs.

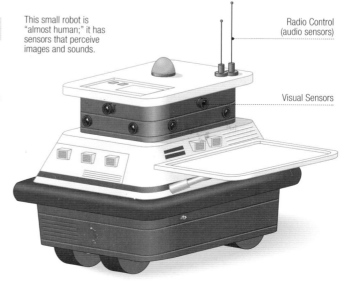

This small robot is "almost human;" it has sensors that perceive images and sounds.

Radio Control (audio sensors)

Visual Sensors

SOME AUTOMATED TASKS

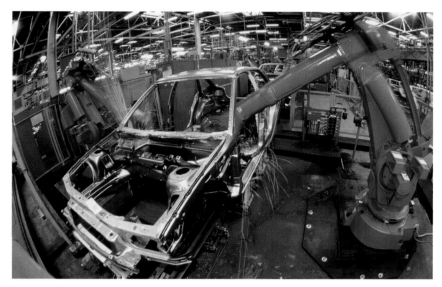

Modern automobile plants are entirely automated.

In addition to the automobile industry, there are other industries involving metal and mechanics where robots perform duties of **welding**, **painting**, **adjusting**, and **assembling**. These tasks are difficult and repetitive, and machines can do them more consistently than humans. There are specialized robots for **deactivation** of explosives, and others that perform **underwater** tasks at depths where no humans can work. Overall, robots are used for particularly dangerous, demanding, and repetitive tasks.

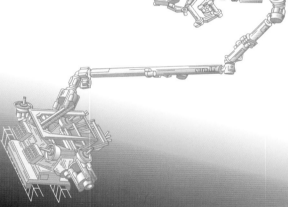

Articulated arm developed by NASA for working on space stations.

SPACE ROBOTS

One of the most impressive tasks that robots carry out is **loading** and **unloading** NASA's space shuttles. This involves the use of a huge **articulated arm** that can move items weighing several tons with total precision. This arm is remarkable not only for its strength, but also for the delicacy with which it manipulates precision instruments that cost many millions of dollars.

VIRTUAL REALITY AND ARTIFICIAL INTELLIGENCE

The Sony dog "obeys" various commands and barks softly when its batteries get weak or when no one pays any attention to it.

Some computer programs create very realistic environments in which a person, generally equipped with special **gloves** and a **visor** that has fine sensors, can move and **interact** as if in a real world. These are **virtual reality** programs. They are used in some games, but they also have important applications to **scientific research**. A large part of the efforts that go into the development of computers with **artificial intelligence** focuses on the field of virtual reality.

AN "ALMOST INTELLIGENT" DOG

In the last few years of the twentieth century, the Japanese company Sony developed a small robot that looked like a dog. It was named AIBO, and it had a **microprocessor** that incorporated several spectacular advances in artificial intelligence, thanks to **genetic algorithms**, a series of mathematical, logical, and ordered operations capable of changing through trial and error.

Introduction

Machines
and Tools

The Steam
Engine

Internal
Combustion
Engines

Energy

Electricity
Production

Electric Motors

Mining

Metallurgy

The Chemical
Industry

Construction
Materials

Public Works

Transportation
Vehicles

Imaging

Electronics

Computer
Science

Robotics

Index